Élisée Reclus

HISTORIAN OF NATURE

BY

GARY S. DUNBAR

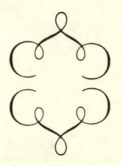

ARCHON BOOKS
1978

© GARY S. DUNBAR 1978

First published 1978 as an Archon Book,

an imprint of THE SHOE STRING PRESS, INC.

Hamden, Connecticut 06514

Library of Congress Cataloging in Publication Data

Dunbar, Gary S
    Élisée Reclus, historian of nature.

    Includes bibliographies and index.
    1. Reclus, Élisée, 1830-1905.
    2. Geographers—France—Biography. I. Title.
G69.R32D86   910'.92'4 [B]   78-17346
ISBN 0-208-01746-1

*This book is
dedicated with love and respect
to my parents,
Esther Seamans Dunbar
and
Alvin Robert Dunbar,
very young contemporaries of
Elisée Reclus*

Contents

Illustrations

Preface

Elisée Reclus (1830-1905) was one of the most prolific geographers of all time—perhaps the most prolific (but who would count all his words and those of rivals in order to substantiate or refute such a claim?)—and for one who is not a native speaker of French to attempt his biography is surely an exercise in what my late friend John K. Wright would call "foolrushery."[1] Not only would it be foolhardy but perhaps also unnecessary, for biographies of Reclus already exist, and, furthermore, the best of these biographical works were written by people who knew Elisée Reclus intimately. Why, then, a new biography? I feel that a new work is necessary in order to introduce Reclus to a new generation, particularly in the English-speaking world, because relatively little material on his life and work has been available in English. I have reviewed almost all of the published sources used by earlier writers, and I have seen much material, especially unpublished sources, which they did not employ. Interest in Reclus, in France and abroad, seems to be growing at the present time, especially among the youth who are constantly rediscovering Reclus's anarchist activities. His name is still being invoked to gain support for radical causes. For example, I have an "Anarchist-Revolutionary Calendar" for 1970, published in Chicago, which includes a fine woodcut of "Elisée Reclus, Anarchist-Geographer" on the April page and mentions such "revolutionary" events as the San Francisco earthquake and the ambush of Bonnie and Clyde!

I must admit that I share the trait common to nearly all biographers of being sympathetic to their subjects. To me Elisée Reclus emerges as a wholly admirable person, almost an Olympian figure, and although I do not think that it would be difficult to persuade readers to adopt a similar view, my aim is neither to glorify Reclus nor to vilify those who held contrary views. It has been suggested that there might be an inherent contradiction in the term "anarchist-geographer" because geographers, like all other scientists, are supposed to be politically neutral. But value-free, or value-neuter, geography may not be possible, as Kenneth Rexroth has recently shown,[2] and geographers should not be surprised or embarrassed to find that their values and prejudices are reflected in their methods and data. I hope to show that geography and anarchism can be a logical combination; at least they were for Elisée Reclus and his colleague Peter Kropotkin. Although I believe that an understanding of Reclus's anarchism is necessary to an understanding of his geography (and the reverse is also true), I intend to tread lightly in discussing his political views, partly because they have already been adequately treated by such scholars as Max Nettlau and Jean Maitron, and partly because it is a virtually impossible task for someone like me to plumb the arcane mysteries of nineteenth-century anarchism. As a geographer I intend to stick quite close to my last.

The professionalization of the academic disciplines, including geography, took place in the last quarter of the nineteenth century and the first quarter of the twentieth. This period saw the proliferation of university chairs and departments, advanced degree programs, professional societies, and journals. Thenceforth, a person was not accepted as a "professional" unless he had the necessary academic credentials, usually the doctorate. In the professionalization of geography, as in so many fields, Germany was clearly the bellwether, as she had created the first truly modern universities. Carl Ritter has been called one of the fathers of modern geography, along with Alexander von Humboldt, because of his long tenure as the first professor of geography in the first modern university, the University of Berlin, but the real growth of geography as an academic discipline came after his death in 1859. Among Ritter's students were not only Germans who later became important in the professionalization of geography, such as Ferdinand von Richthofen, but also foreign students such as Peter Semënov, Arnold Guyot, and Elisée Reclus.

[10]

There has been considerable interest in recent years in the professionalization or institutionalization of the social sciences, including geography and history, and it would seem that France has received the fullest coverage, thanks especially to North American scholars. Terry Clark has described the rise of the social sciences in France; Benjamin Harrison and William Keylor have treated the emergence of the French historical profession; and Vincent Berdoulay has produced a masterful survey of French geography in the period 1870-1914.[3] It may not be inappropriate, then, for me to offer this biography of Elisée Reclus, the leading figure in what might be called the protoprofessional period of French geography.

Although he was a prolific writer with perhaps the largest readership of any modern geographer, Reclus could not have initiated the professionalization of geography in France, for he did not have the doctorate that was necessary for becoming a professor in a French university. He also lived outside of France during the major part of his adult life. The title of "The Father of Modern Geography in France" has been conferred on Paul Vidal de la Blache (1845-1918), a Sorbonne professor who had been trained in classical history, but Vidal and his colleagues readily admitted their debt to Reclus, whose work had paved the way for the professional or academic geographers. "Reclus undertook the conversion of the French to geography," said his cousin Franz Schrader, himself an outstanding nonacademic geographer of the Vidalian epoch.[4]

In employing the title "Historian of Nature," which I borrowed from a review of one of Reclus's works by Georges de Nouvion,[5] I am showing that I shall emphasize the scientific side of Elisée Reclus's life and, more precisely, his rôle as a scientific popularizer or "literary" geographer. I should like to aid, but not push, the reader to make for himself the discovery that Reclus's message is ever-new and most definitely relevant to the 1970s, in much the same manner as the writings of Lewis Mumford, who is a direct spiritual descendant of Elisée Reclus. My book is not offered as a memorial to Reclus—such hagiography would have been repugnant to him—but it is presented as a simple narrative of the efforts of one man, who happened to be a geographer, to come to grips with the overarching issues of his day, which the reader can easily recognize as some of the leading problems of our own day.

Acknowledgments

I have often been asked how and when I first became interested in writing the life of Elisée Reclus. I honestly cannot remember when the idea came to me or what my original motives were. I might have been introduced to Reclus in an inspiring course taught by Professor Robert West at Louisiana State University in 1953. I was intrigued by the life of this man who was outside the mainstream of academic geography in his day. Apart from taking notes on some published works by and about Reclus, I did not engage in serious research into his life until 1966, when I spent twelve weeks in Edinburgh on home leave from Ahmadu Bello University in Nigeria. I had chosen Edinburgh specifically because of Elisée Reclus's connection with Patrick Geddes. The Geddes Papers in the National Library of Scotland yielded considerable Reclus material, and Dr. and Mrs. Arthur Geddes were most helpful. (Dr. Geddes, who died in 1968, was Patrick Geddes's son, and Mrs. Geddes is a great-granddaughter of Elisée Reclus.) In 1967, on my return to the United States from Nigeria, I was able to visit places with Reclus associations in southwest France, especially Sainte-Foy-la-Grande and Domme, and also Amsterdam, where I used the unrivalled collections on nineteenth-century socialism and anarchism in the Internationaal Instituut voor Sociale Geschiedenis. In Los Angeles, from 1967 to the present, I have been given every form of encouragement by the University of California to pursue my work. In the spring of 1971 a grant from the Research Committee of the Academic Senate enabled me to spend a month in Paris and nine days

[12]

in Brussels. Also in 1971 I was fortunate in obtaining the notes that Professor George Engerrand, a younger colleague of Elisée Reclus in Brussels from 1898 to 1905, made for a projected biography of Reclus before his own death in 1961.

As for individuals who have helped me along the way, I must mention specifically the members of the Reclus family, who have carried on the family traditions of generosity and benevolence, manifest especially in their contacts with an American geographer whose French is faulty and who has frequently displayed ignorance and insensitivity to French ways: Mrs. Jeannie Collin Geddes of Edinburgh; M. Jacques Reclus of Paris, a great-nephew of Elisée Reclus; the late M. Michel Reclus of Paris and Larmor-Baden, Jacques' brother; Mme. Louise Rapacka of Paris, a great-granddaughter of Elisée Reclus; and M. Jacques Régnier of Marseille, a grandson of Reclus. In Sainte-Foy-la-Grande the late M. Jean Corriger gave me invaluable aid, now fortunately continued by his wife, and in Domme M. Robert Degouy provided welcome assistance. In Paris I received the anonymous but nevertheless generous assistance of members of the staffs of the Bibliothèque Nationale and the Archives Nationales, and I should like to acknowledge the prompt and kind attention paid me by Mme. Denise Fauvel-Rouif and her staff in the Institut français d'histoire sociale. In Brussels I was the recipient of numerous acts of generosity on the part of M. F. Sartorius and Mme. Andrée Despy-Meyer of the Service des Archives of the Université Libre; Professor François Stockmans of the Académie Royale des Sciences, des Lettres et des Beaux-Arts de Belgique; and M. Antoine de Smet of the Bibliothèque Royale. Dr. Rudolf de Jong of the Internationaal Instituut voor Sociale Geschiedenis in Amsterdam has been very helpful. Mrs. Anita Engerrand Gafford of Denton, Texas, was exceedingly generous in providing me with her late father's notes, and I should also like to acknowledge the previous assistance of her late brother, Professor Jean Jacques Engerrand of Kent, Ohio. Dr. Vincent Berdoulay of the University of Ottawa has been helpful in ways too numerous to mention. Professor and Mrs. Ronald F. Lockmann were my genial hosts for a week in New Orleans, and Professor Lockmann subsequently rendered valuable bibliographical assistance.

I have been fortunate in securing the services of UCLA students in the translation of materials from languages in which I have little or no proficiency: Mrs. Helge Prosak, Mrs. Lotte Ehrlich, and Mr. Josef Koch (German); Mrs. Consuelo Wager Dutschke (Italian); Mr. Ilya Talev (Russian); Mr. Carl Dahlman (Swedish); and Mrs. Annette L. Creasy (Dutch). Mlle. Jeanne-Christine Gué was responsible for a particularly difficult translation. My translations from French are sometimes free and sometimes literal, depending on what I think is appropriate. Following convention, I have not used accent marks on capital letters, as in the name "Elisée" itself.

Others who provided information or encouragement at some point include Dame Rebecca West, Professor David Lowenthal, and Professor William Mead, all of London; Professor Aimé Perpillou, Professor Philippe Pinchemel, Professor Paul Claval, and Dr. Béatrice Giblin of Paris; Professor P. Tucoo-Chala and Mme. Michelle Imbert of Pau; Dr. Jacqueline Desbarats of Pau and Pinang; M. Jacques Duzer of Dijon; Dr. Gerhard Engelmann of Potsdam; Mr. Lewis Mumford of Amenia, New York; Dr. Meredith F. Burrill of Chevy Chase, Maryland; Mrs. Crystal Ishill Mendelsohn of Columbus, Ohio; Professors Marvin W. Mikesell and William D. Pattison of Chicago; Professor Donald D. Brand of Austin, Texas; Mr. Sidney L. Villeré of New Orleans; Mr. Kenneth Rexroth of Santa Barbara, California; Professor Geoffrey Martin of New Haven, Connecticut; Professor Edward J. Miles of Burlington, Vermont; Professors Clarence J. Glacken and David J. M. Hooson of Berkeley, California; and Professor Paul C. Smith of Los Angeles. Mr. Noël Diaz of UCLA drew the maps, and the manuscript was typed by Dexter Pointer.

G. S. D.

University of California
Los Angeles

Chapter One

The Awakening

"Reclus is not a Frenchman—I am sure of that, though he speaks French very well and fast. He is too proud for a Frenchman. He may be a Gaul."[1] Thus wrote John Bigelow, former American minister to France, to his friend William Huntington, Paris correspondent of the New York *Tribune*, in 1868. Bigelow very well knew that Elisée Reclus was indeed a Frenchman, but he was saying that Reclus was not an ordinary sort of person, not the ordinary Frenchman one might meet in Paris. There was something in Reclus's character that was perhaps reminiscent of ancient Gaul before the time of the Roman and Germanic invasions. Although the mature Elisée scorned narrow chauvinism, he always remained a French patriot in the best sense, loyal to the land if not to the government.

The Reclus family derived from Gascony or Aquitaine, the southwestern part of France, which has always been a thoroughfare of peoples and ideas. In the Périgord, not far to the east of Elisée Reclus's birthplace, are such places as les Eyzies-de-Tayac and Lascaux, among the most important late Paleolithic archeological sites in all Europe. In such an ancient region genealogy is interminable, and the ethnic skein is impossible to unravel. It was perhaps, then, no great exaggeration when Paul Reclus, Elisée's nephew, described the family links with the uttermost reaches of Europe and even Asia and Africa. "By the mixture of bloods the Reclus thus undoubtedly had cousins thirty times removed in Scandinavia, Mauritania, Great Britain, Spain, and elsewhere. More distant relatives might perhaps be found in Germany, Russia and as far as Mongolia on the one hand, and as far as Italy,

[15]

Greece, and Syria on the other."[2] Paul was not, of course, basing this statement on any precise genealogical information, and it would apply not only to the members of the Reclus family but to all Gascons, or all Frenchmen, or, indeed, all Europeans. He was simply saying, in a manner typical of the nineteenth century, that the Reclus were part of the universal brotherhood and did not subscribe to the notion of racial "purity" which has proved to be so iniquitous in so many parts of the world.

The Reclus and allied families have inhabited southwestern France since time immemorial. The name, which is the same as the English word "recluse," is somewhat ironic because the Reclus have been anything but recluses. The name appears in Sainte-Foy-la-Grande as early as the beginning of the thirteenth century. As converts to Protestantism, presumably in the sixteenth century, they suffered from religious persecution, particularly in that period in the eighteenth century when the Reformed Church was known as the "Church of the Desert."[3] This period of repression had the result of toughening the spirits of the French Protestants, which may account for their disproportionately great representation among leaders of the professions in nineteenth- and twentieth-century France. Out of this crucible came Elisée's great-grandfather, Jacques Reclus (1710-1796), who was described in his marriage contract by a facetious notary as a "cooper, heretic, and scholar." His son Isaïe, Elisée's great-uncle, is said to have recited verses of The Aeneid while plowing his fields. One can see that the traditions of learning and of heresy showed up in the family at least two generations before they became known for these qualities all over France.[4]

Our story properly begins with Elisée's father, Jacques Reclus (1796-1882), the patriarch of the *tribu*, as his numerous progeny were known in the later nineteenth century.[5] Born in Le Fleix, a small village on the Dordogne River where his father was a rather prosperous farmer, Jacques was educated for the ministry at the Protestant university at Montauban. He was consecrated at Nîmes in 1821 and served briefly as an assistant pastor there, but in 1822 he returned to his home district to preach, first at La Roche-Chalais, where he was married, and then at Montcaret-Le Fleix. While serving in the latter post, he lived in Sainte-Foy-la-Grande, and it was there that Elisée and his older brother and sister were born. A spirit of revival was sweeping through the Reformed Church in the late 1820s, and the young pastor was won over by the new ideas. The "théologie du Réveil" ("Awakening") was introduced into the Sainte-Foy area in August 1828 by a Swiss pastor, Alexandre Henriquet. In May 1829 Henriquet was rejected by the local consistory, and Jacques Reclus was the only person voting against this decision. Early in 1831 Reclus declared that his doctrines "were other than those which he had on his entry into the Church," and he then resigned from his secure position and became the pastor of a "dissident" community (Free Church) in Orthez and environs in the Basses-Pyrénées, where he remained for more than half a century, until his death.[6]

A man of firm belief and undying faith, Jacques Reclus lived the Christian life so thoroughly that he has been described as a true communist or anarchist. Little concerned with material goods and not at all with egotism, Jacques became a living legend in the Orthez region for his numerous acts of generosity and altruism. There are many stories to illustrate these qualities. For example, when he learned that poor folk were stealing his potatoes he blamed himself for not being mindful of their needs and then set some potatoes out at the edge of the field for them to share. He would give away not only his surplus but also money and clothing that were needed in his own household. Not only did he give away his old frockcoats, but one day in January he gave away a new one. When he was an old man, he saw a robust young peasant burying a diseased horse, and he insisted on taking over the job himself because he was afraid that the young man might contract a fatal illness. These stories may be apocryphal, but they could well be true of this saintly man who throughout his long life practiced an extremely consistent and literal belief in the Holy Scriptures. His saintliness his children could revere, and his inflexibility they could not directly combat. It was his hope that his two eldest sons would follow him in the ministry, but although they took university training in theology, they rebelled against religious strictures and dogma, declared themselves to be atheists, and yet displayed a religious zeal and missionary spirit to the end of their days. The sons did not extinguish the father's faith; it was simply transmuted.[7]

Of equal importance to the formation of the children's character was their mother, Marguerite Zéline Trigant, born in 1805 into a rather well-to-do family at La Roche-Chalais on the Dronne River about forty-five kilometers to the northwest of Sainte-Foy. She married the young pastor of La Roche-Chalais in 1824, and their numerous children, eleven of whom lived to full maturity, were born in Sainte-Foy and Orthez. To supplement her husband's meager income, she conducted schools in Castétarbes and Orthez where more than three generations of local girls were educated. It is said that she arose at 3:00 A.M. in order to prepare lessons. When she was almost seventy years old, she realized that there was a gap in her knowledge—in physics—and so she set out to master the subject in order to inform her pupils. Elisée always remained close to his mother, spiritually if not physically, although she was always so busy as housewife and teacher that she was not able to lavish attention on him. He described her as "admirably zealous, but of another sort than her husband," and another observer made the same point when he said, "She was a saint according to humanity, as her husband was a saint according to Protestant orthodoxy." It was said that she had a greater wanderlust than any of her children, although her entire life was spent in southwestern France. After her husband's death she went back to Sainte-Foy to live with her daughter, Zéline Faure, and there she died in 1887.[8]

To this remarkable couple were born numerous progeny. The eldest, Suzi, died at age twenty, and the youngest, Anna, died in her first year, but between these two unfortunates were born eleven children, six girls and five boys, all of whom lived into old age and made, collectively, a considerable mark. After Elisée and his older brother, Elie (born Jean Pierre Michel on 16 June 1827), came, in chronological order, Loïs (1831-1910), Marie (1832-1918), Louise (1835-1917), Onésime (1837-1916), Zéline (1838-1911), Noémi (1841-1915), Armand (1843-1927), Ioanna (1845-1937), and Paul (1847-1914).[9] It would be tempting to delve deeply into the lives of each of these interesting individuals, but here it will only be possible to treat them in their relations with their brother Elisée.

Elisée's birthplace, Sainte-Foy-la-Grande, was, and remains, a small village on the Dordogne River some sixty-five kilometers east of Bordeaux. It was founded in 1255 as a *bastide* (fortified town) by the Count of Toulouse and passed into English hands some two dozen years later. Largely rebuilt in the last two centuries, Sainte-Foy today still has many old buildings, including the one in which Elisée was

Photograph of the five Reclus brothers taken by Nadar (*père* or *fils?*). Undated (probably 1893). From left: Paul, Elisée, Elie, Onésime, and Armand. From copy of photograph that the late M. Jean Corriger of Sainte-Foy-la-Grande had borrowed from Mme. Dr. L. Bouny of Brussels.

born. The population, 3186 in 1968, has remained remarkably stable in the last two centuries. In 1780 it was estimated to be 3000, and in 1868 it was 4033. In the nineteenth century it was noteworthy for its large Protestant minority—it has been called "The Geneva of the Southwest"—and for its numerous schools. Pastor Reclus did some teaching in the college (high school) that Pastor Celestin Bourgade founded in 1824 and which Elisée attended in the 1840s. Sainte-Foy was also the home of the Broca family, relatives of the Reclus; one member, Paul Broca (1824-1880), became a well known anatomist and anthropologist.[10]

Elisée Reclus was born on 15 March 1830 at 4:30 P.M. His name was recorded as Jacques Elisée, but he did not use his first name in his adult life. Some writers have said that his full name was Jean Jacques Elisée, but I have never seen his name officially recorded in that way. His older brother had been christened Jean Pierre Michel, but he was always known as Elie, after his godfather, the Duke Elie Decazes, a distant relative on his mother's side and at whose home Pastor Reclus was once employed as librarian. The names "Elisée" and "Elie" (English equivalents are Elisha and Elijah, respectively) are names that were used quite frequently by French Protestants in the nineteenth century. Because of the similarity of their names, Elisée and Elie have sometimes been confused by bibliographers, usually by attributing one or more of Elie's writings to his more prolific brother.[11]

The Reclus family did not long remain intact in Sainte-Foy after Elisée's birth, for in the very next year Jacques Reclus resigned his position to take up a pastorate in the Free Church in the Orthez area. Elisée stayed with his maternal grandparents in La Roche-Chalais and joined the family in Castétarbes several years later, in 1838. Thus his earliest memories were not of his parents' home but of the grandparents'. Reclus's first geography lesson, according to his own testimony, came when his grandfather told him of the limitless Sahara, "which never ends but always begins again." It is said that his vivid childhood memory of the Dronne River at La Roche-Chalais inspired his popular book, *Histoire d'un ruisseau*, published in 1869.[12]

Jacques Reclus wanted a strict religious and classical education for his children. He was an admirer of Count Zinzendorf and the Moravian Brothers, and so it happened that he sent his two oldest children, Suzi and Elie, off to the Moravian school in Neuwied, Germany, in

1840. Elie remained in Neuwied until 1843, and Elisée, who was sent there in 1842, stayed until 1844. The choice of distant Neuwied for the education of the children was truly remarkable. The Moravian school—actually two schools, one for girls and one for boys—was founded in 1756, and in the first half century of its existence most of its pupils came from Switzerland. According to Max Nettlau, the school was not a *Gymnasium* but a sort of *Mittelschule*. In the 1830s and 1840s there were large numbers of English students at Neuwied, including George Meredith (1828-1909), who was later to become a prolific novelist. Meredith's years at Neuwied, 1842-1844, coincided with Elisée's, but, strangely, there is no record of any connection between the two at that time or later in life. In his obituary of Elie, published in 1905, Elisée mentioned that Elie had been a friend of George Meredith in school. Although I cannot find any connection between Elisée Reclus and George Meredith in their mature years, George Engerrand, who was close to Reclus in the period 1898-1905, claimed that Elisée liked to visit Meredith in England, even though he could do it only rarely. Otherwise the records are mute.

The Reclus brothers not only learned German and Latin in school, but they picked up English and Dutch from their classmates. The teachers were interested in nature study and often took the children on hikes along the Rhine. This experience definitely reinforced the love of nature already present in the Reclus brothers. It is said that Meredith's passion for long walks and hill-climbing probably developed at Neuwied. These early impressions of the Rhine had a profound effect on Elisée Reclus. He was to retain an idealized view of the Rhineland of his youth, which he later contrasted with the polluted industrial landscape of the late nineteenth century.

Altogether, the Neuwied experience was apparently a happy one for the Reclus brothers, although Elie had been severely hazed by his fellow students when he matriculated, and although they lived a stark existence far from home. Of inestimable value to their later work was their early exposure to the English and German languages. Elisée liked Neuwied well enough that he returned to teach French for a year (1849-1850), but in his old age he castigated the school and the Moravians in the strongest terms. In his obituary of Elie, Elisée deplored the "devouring zeal" of the Moravian Brothers, whose lives were "regulated in advance by a disgusting ritual of infantile practices and conventional lies." "The director of the two establishments for girls and boys was a cowardly man, happy to fawn servilely on those students whom he knew to be rich, and to scoff and sneer at those who he knew were poor." The Reclus brothers were probably the poorest of all. A measure of their poverty was the fact that their mother calculated that since postage cost thirty-eight sous she could only afford to send her sons a letter once every two months.[13]

After two years at Neuwied Elisée returned to his natal village to complete his secondary education in the Protestant College of Sainte-Foy. He was awarded the diploma of Bachelor of Letters from the University of France (the state educational system) on 14 November 1848.[14] (It should be made clear to those who are unfamiliar with French practices that this diploma was not like a Bachelor's degree from an English or American university, despite the name. The French *baccalauréat* represents the end of preparation for the university. It is akin to sixth-form preparation in a British school and therefore much more advanced than high-school training in the United States.) While studying in Sainte-Foy, Elisée and Elie lived with their maternal aunt and uncle, the lawyer Chaucherie. The aunt disciplined the boys rather harshly, and the uncle antagonized them with his preoccupation with materialistic concerns. Nettlau has suggested that Elisée might have become interested in socialistic writings at this time out of opposition to his uncle. Contact with the advanced social ideas of the day (*e. g.*, the ideas of Saint-Simon, Comte, Fourier, and others) was apparently provided the Reclus brothers by a militant socialist named Dupuis who had been a worker in Paris before settling in Sainte-Foy in 1837. There is no evidence, however, that the brothers were converted at this time, and their contemporary ideas about the revolutionary events of 1848 were not recorded. They were both still aiming toward a career in the Protestant ministry. After his completion of the baccalaureate in the College of Sainte-Foy, Elie went to Geneva in 1847 to study theology. He read demonology more avidly than the traditional theology, suffered an unrequited love affair, and was generally penniless and lonely during his stay in Geneva. In 1848 he returned to France to join Elisée at the Protestant university of Montauban. During most of 1849 Elie, Elisée, and their old friend from school days in Sainte-Foy, Edouard Grimard, lived in a little house on a hill some four kilometers from Montauban.

There they read Proudhon and other unorthodox scriptures. After taking an unauthorized leave from their studies to travel on foot to see the Mediterranean, where, characteristically, Elisée became so carried away by the view that he bit his more stoical brother on the shoulder "until the blood came," the three young men were reprimanded by the university officials, and they then decided to leave Montauban and pursue their studies elsewhere. Grimard later became a botanist, aided Elisée with the botanical portion of *La Terre* in 1868, and died in 1908. Upon leaving Montauban in 1849 Elie went to the University of Strasbourg, where he completed his theological studies in 1851 with a thesis on "The Principle of Authority." He was then ordained as a minister, but he promptly resigned his commission. [15]

Elisée returned to Neuwied in late 1849 and became an assistant master in his old school, teaching French for two hours a day. Desiring to continue his university studies in theology, he thought of enrolling in either Leipzig or Halle, but on the advice of one of the Moravian Brothers named Geller he went to Berlin instead. It is not true, as many authors have said, that Reclus was attracted to Berlin because he wanted to study under the great geographer Carl Ritter. It is clear that his original motive was to study theology, and there is no proof that he had ever even heard of Ritter before going to Berlin.

Elisée Reclus attended courses at the University of Berlin for about a half year beginning in February 1851. At the same time he gained a meager income by giving French lessons. Some authors have erroneously stated that Reclus studied in Berlin for as much as two years, from 1849 to 1851. In an early letter to his parents he said that theology was less well represented in the University of Berlin than other fields of study and that theology students were free to take any courses they wanted. Among the 120 professors in the University he named five as outstanding—two in theology, one in political economy, one in the history of medicine, and Carl Ritter in geography. At that time Ritter was seventy-one years old and was teaching large classes of "comparative" geography, a very general world human geography. There is no indication that Reclus ever had any intimate contact with Ritter. I can find no evidence that there was any correspondence or even a conversation between the two. Although Reclus visited several acquaintances when he returned to Berlin in 1859, there is no evidence that he visited Ritter, who died the following month. Reclus always alluded very warmly to his old mentor, and his eulogy of Ritter in the *Revue germanique* in 1859 is one of the most insightful portraits of Carl Ritter in existence. Also, I have no proof of any contact between Reclus and Alexander von Humboldt, whom Reclus admired greatly and who died in the same year as Ritter.

Reclus's formal education ended in the summer of 1851 when he left the University of Berlin and went home. On the eighth of July he announced that he was ending his theological studies, but an Exmatrikel, the official notice issued when one leaves the university, was not recorded in his case, so that the otherwise customary facts about his studies are not known. It may be that he intended to return in the near future.[16] In any event, he left Berlin and went to Strasbourg, where Elie had just completed his work, and the two brothers, accompanied by a dog, then walked from Strasbourg to Orthez. This remarkable peregrination has been distorted by some and has even been given symbolic value. It has been said that both Elisée and Elie walked to Berlin in 1849 to study under Ritter, but no part of that statement is true: Elie did not go to Berlin, and Elisée went in 1851 but not because of Ritter. Patrick Geddes, who did not like to have his flights of fancy impeded by strict adherence to facts, tells the story in this way:

> The world-experience which made Elisée Reclus the supreme geographer of his day grew out of his youthful long and adventurous tramps with his brother Elie . . . in crossing yearly between their home at Montauban near the Western Pyrenees, and their school at Neuwied, well up the Rhine. This sort of touring, at its simplest, is now reviving among the youth of Germany, in its poverty, much as in the old days of wandering students and apprentices.[17]

The brothers reached Montauban three weeks after leaving Strasbourg. It is possible that, during this visit to Montauban, Elisée wrote an essay, "Développement de la liberté dans le monde," which has been called his first anarchist writing. (This paper was almost discarded by Elisée late in life, but his sister Louise gave it to Clara Mesnil, and it was published for the first time in 1925.)[18] After several more days on the road Elisée and Elie reached Orthez, only to find the family full of grief because baby Anna had just died after a brief illness. The brothers' activities and plans during the autumn of 1851 are not known, but there occurred in early December an event which trans-

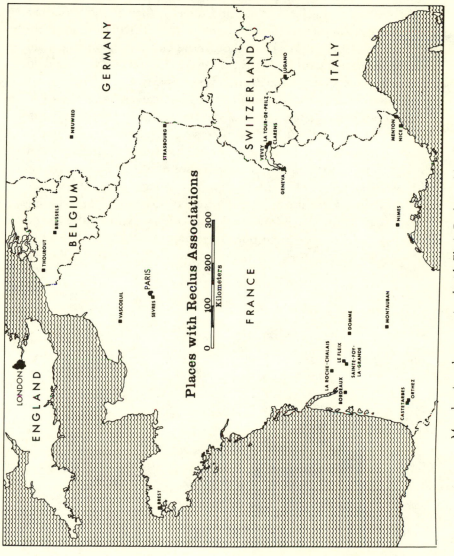

Map showing places associated with Elisée Reclus and his family.

formed their lives; in fact, one might say it radicalized them. The young men were staunch supporters of the Republic which had been established in 1848, and they were profoundly shocked when Napoleon III reestablished the Empire with his coup d'état of 2 December 1851. After the Reclus brothers and some of their friends made an unsuccessful attempt to take over the Orthez Town Hall as a revolutionary gesture, it looked as though the brothers might face a jail sentence, so their mother gave them the very large sum of five hundred francs in order that they might leave the country until things cooled off. The brothers were not officially exiled or even, so far as I can determine, officially reprimanded, but they simply chose to leave in advance of any such strictures. It is possible that they had thought of going to England even before the coup, but the town-hall incident hastened their departure. In any event, we find the brothers leaving Orthez in late December and arriving in London on New Year's Day 1852. Thus began Elisée's longest period of overseas travel—lasting more than five and a half years—a period which provided him with the desire and experience to become a geographer.[19]

Chapter Two

The New World

The Reclus brothers' self-exile occurred at a most convenient time and fits the traditional German notion of *Wanderjahre*, the years of travel taken by young men after they finish their *Lehrjahre*.[1] Characteristically, Elie traveled less far and returned sooner to France; his impetuous younger brother was to spend several years in the New World, always trying to persuade Elie to join him, before returning to his homeland in 1857.

The brothers arrived in London on New Year's Day 1852. The money which their mother had given them was apparently largely consumed in traveling to London from Orthez, because they soon began to suffer from inadequate housing, clothing, and food. Their sister Loïs previously had voiced her disappointment with London when she visited the city en route to a tutoring job in Scotland. Elisée told her that London was a great center of activity and that the English had done more than any other people to make all men brothers, but he revised his opinion on his first visit. He and Elie were treated with suspicion by Londoners because they were not dressed like "gentlemen" and because French refugees of that period were all thought to be socialists. Indeed, the brothers thought of themselves as socialists by the time they reached London, and Elisée was to attend lectures by Louis Blanc and other socialist speakers there. This was a time of great experimentation for the young men, for—if we can take literally a cryptic passage in a letter Elisée wrote to Elie—they claimed to be the only Swedenborgians and homeopaths in all of London. I can find no further indications of these latter interests, and they may represent simply evanescent youthful aberrations. Among the French people Elisée met in London was the impoverished L'Herminez family, including young Rosalie ("Fanny"), who was to become his second wife in 1870.[2]

The brothers were finally driven to that last resort of educated but penniless young men—teaching. Elie was fortunate in becoming the tutor for a most remarkable Anglo-Irish family, the Fairfields, of whom one of the sons, Arthur, became the father of the outstanding twentieth-century writer Rebecca West. Dame Rebecca, now an octogenarian, still acknowledges her great intellectual debt to Elie Reclus, although she has confused him with Elisée.[3]

Elisée Reclus gave French lessons for a brief time to Richard Heath, scarcely younger than himself, and this encounter led to a warm life-long friendship. After meeting again thirty years later Heath and Reclus kept up an active correspondence until the latter's death. The correspondence is remarkable in that Heath, who wrote many books on religious and social questions, had ideas that were very close to Elisée's, although his explanations were always couched in a Christian framework. Reclus thought that Heath was misguided, but he did not treat his ideas with the withering criticism that he would have used against those who were not his friends.[4]

During his stay in London Elisée visited the huge globe that James Wyld, "Geographer to the Queen," had erected in Leicester Square as a money-making scheme in connection with the Great Exhibition of 1851. Wyld's "monster globe"—a globe of sixty feet diameter with the continents portrayed on the inner surface for ease of viewing, much like the earlier "géoramas" in Paris—was indelibly impressed on Elisée Reclus's mind, for it was the inspiration for the large globes that he proposed more than forty years later.[5]

Later in 1852 Elisée was employed as an estate manager in Ireland, a job for which he had the inclination but no experience. He was employed to give advice on the cultivation of a neglected farm of eighty-two hectares (about 200 acres) in County Wicklow about fifty kilometers south of Dublin. It was on a visit to western Ireland that he conceived the plan for his first major geographical work, La Terre. He gathered data in his head and in notebooks during his residence in the Americas and for a decade thereafter until the publication of the first volume of La Terre late in 1867. In the preface to La Terre he told how he sketched the basic plan for the work while lying on the grass near the ruins of an old castle on a hill overlooking the Shannon River in the autumn of 1852. The work has thus commenced, he said, not in the silence of a study but in free nature. Reclus emphasized original field study and sometimes denigrated the value of books, and yet his major works are noteworthy for the paucity of personal observations. It was wide reading rather than extensive travel that informed his books.[6]

Next we find Elisée in Liverpool offering to work his way across the Atlantic on steamers bound for New York, but he ended up on a sailing ship which took him to New Orleans instead. He recorded the name of the ship as the *John Howell*, but my cursory search has failed to find such a name. I did, however, find similar names, such as *John Garrow* and *John Harward*, and it is possible that Reclus made an error in transcribing the name. The *John Garrow*, for example, made regular trips between Liverpool and New Orleans, carrying salt to Louisiana and returning with cotton and flour. It is not known when Elisée arrived in New Orleans, but the date is likely sometime in the first half of 1853. After working briefly as a dockworker and narrowly escaping serious injury, he turned to the sort of work he knew best, and for the rest of his stay in Louisiana, about two and a half years, he was a tutor to the children of the Fortier family on their sugar plantation in St. James Parish, on the other side of the Mississippi River from New Orleans and about fifty miles upstream.[7]

Septime Fortier had married Félicité Emma Aime, daughter of Valcour Aime, the well-known Louisiana planter whose estate, the St. James Refinery plantation, was known as "Le Petit Versailles de Louisiane." In 1845 Aime built a great house, Félicité (also called by the anglicized form, Felicity), for his daughter and son-in-law between the Refinery plantation and Bon Séjour (or Oak Alley), the splendid plantation home of his brother-in-law, Jacques Telesphore Roman. Aime's home no longer exists, but Felicity and Oak Alley survive, the latter acclaimed as one of Louisiana's great tourist attractions because of the magnificent avenue of live oaks leading to the house from the road and river.

Elisée lived at Felicity and instructed several of the numerous Fortier children in basic subjects. The oldest were Anna (born 1842) and Michel (born 1843), who were thirteen and twelve years old, respectively, when Reclus departed from Louisiana. The children liked their tutor very much—the eldest ("a girl with large eyes") perhaps too much. Almost fifty years later Elisée told Clara Mesnil that Anna had fallen in love with him and that, although he did not reciprocate, he had a very deep affection for her. He did not actually "flee" from this awkward situation, as he said to Clara, but it was developing at the time he was planning to leave anyway. There is no evidence that Elisée was ever again in touch with any member of the Fortier family. Anna died a spinster in 1930. She remained unmarried not because of her early crush on Elisée Reclus but, according to the family genealogist, because her fiancé was killed in the Civil War. Michel Fortier was murdered at the polls during the Louisiana gubernatorial election of 1883, and the father, Septime, who had offered to stake Elisée in a plantation enterprise in the Amazon region, died in 1898.[8]

Reclus's best friend in his Louisiana days was the local doctor, Claude J.–B. Lafaye (b. 1816), who had immigrated from Martinique in 1848. Lafaye ("a warm, faithful, courageous, erudite, and eccentric man"—which could have been a description of Reclus himself) nursed Elisée through a bout with yellow fever. It is said that Reclus's first published work was an article in a local newspaper which he signed with the name of Lafaye, but no one has ever found the article. Lafaye tried to persuade Elisée to teach his daughter Blanche, who was then eight or nine years old, but Reclus thought that it would be improper for him to teach other children while he was in the employ of the Fortiers.[9]

Elisée attempted to interest Elie in coming out to America. He specifically recommended Massachusetts but also thought that he might be interested in acquiring land in Texas or even in teaching in one of the Louisiana colleges where Elisée himself had failed in attempts to become a science instructor. Elisée planned to become a farmer, not in the United States but in Latin America. At first he considered Mexico but then soured on the idea when he found that it was just another land of passports and police and Santa Anna another Napoleon III. Instead, he began early in 1855 to concentrate his thoughts on New Granada (Colombia), which he envisioned as a tropical paradise, a place where he could carve out a ten-hectare farm and lead an idyllic life.

While living with the Fortier family Elisée was able to travel throughout much of southern Louisiana, and sometime in 1855 he even visited Chicago. He did not describe his trip to Illinois in any detail but only casually mentioned that he had made a foray into the prairies just west of Chicago. There he described some early precursors of today's "sidewalk" and "suitcase" farmers of the Great Plains. "I have seen farmers transport horses and a reaper by railroad, get out in the middle of the grassland, and immediately launch their team across the high and thick grass; when evening came, the return train took them and their hay back to Chicago." He spoke admiringly of the optimistic spirit of the prairie farmers: "The American does not wish to admit that nature is stronger than he is, and even when he builds a hut, he pretends that this hut is the first of a future Rome." The monotonous regularity of the American rectangular survey system offended Reclus's Gallic sense of landscape aesthetics. "The buyers of these . . . squares are never permitted to deviate from the straight line; true geometers, they construct their roads, erect their cabins, dig their fishponds, and sow their turnips with reference to the meridian or the Equator. Thus the prairies which were formerly so beautiful, with softly undulating contours . . . are today no more than an immense checkerboard."[10]

Reclus found black slavery utterly repugnant, and it probably hastened his departure from Louisiana. He deplored the dehumanizing effects of slavery on the white race, as well as on the black. The following story illustrates this feeling very well:

> One day I was caressing the blonde head of a charming little creole who was all laughter and tenderness, and I asked him, as one ordinarily does with children, if he wanted to grow up [sic—not what he wanted to do when he grew up]. Oh, yes, he told me. And why? To beat the slave woman. The child who expressed this cruel wish was of extreme sweetness but all that he saw proved to him that the privilege of adults was to beat and whip.[11]

Elisée also recoiled at the debased condition of the Indians still surviving in Louisiana. One day, when traveling through the pine woods east of Lake Borgne, he encountered "King Denis, leader of a dozen Indian beggars." "Dirty, hideous, covered with rags. . . he was stretched out under a tree in a state of complete drunkenness."[12]

Although Elisée had been falling away from the church or at least from religious orthodoxy since his school days in Neuwied and Sainte-Foy, it was perhaps his experiences in the New World that caused him to leave the fold, never to return. He must have felt rather isolated in rural, Roman Catholic Louisiana. The atmosphere of America is antimystical, he said, producing "that general atheism of all Yankees, from the Bostonian to the Creole."[13]

And so the great horror that he developed for slavery, the church, and creole chivalry, combined with sheer wanderlust and the romantic notion of becoming a tropical planter, became so overwhelming that he left Louisiana in late 1855 on the steamer *Philadelphia* bound for Panama. His place as tutor to the Fortier children was taken by "a charming young lady from New England."

In letters to his brother and mother before leaving Louisiana, Elisée described his great interest in geography. To Elie and his wife he wrote:

> You know—or rather you don't know—that I have been pregnant for a long time with a geographical *mistouflet* [dial. "a lively child"] that I wish to bring forth in the form of a book; I have already scribbled enough, but that doesn't satisfy me; I also want to see the Andes in order to cast a little of my ink on their immaculate snows.[14]

In a letter to his mother on 13 November 1855 Elisée revealed his plan to go to South America, there to settle perhaps on one of the Colombian or Peruvian tributaries of the Amazon:

> I need . . . especially to gaze upon those Cordilleras which I have dreamed of since infancy and which are so near, on the other side of the Gulf of Mexico . . . For me, seeing the earth is studying it; the only truly serious study that I do is geography, and I believe that it is much more worthwhile to observe nature firsthand than to imagine what it is like while sitting in one's study. No description, no matter how beautiful it might be, can be perfectly true, for it cannot reproduce the life of the landscape, the flowing of water, the rustling of leaves, the song of birds, the perfume of flowers, and the changing shapes of clouds. In order to know, it is necessary to see . . . That is why I wish to see the volcanoes of South America.[15]

Incidentally, in both of these letters Elisée mentioned his interest in vegetarianism. In South America, he said to his mother, "a vegetarian like me makes a delicious meal of manioc and bananas, and, in that way, he can live on three sous a day." So far as I can determine, these are the first mentions of vegetarianism in Reclus's writings. Reclus practiced vegetarianism consistently in his middle and later years, and it was an important part of his ethical code.[16]

Elisée purchased a notebook from a New Orleans stationer in 1854 and filled it with geographical information during the next seven years. Unfortunately, he did not record personal observations but only excerpts from various publications. Many of these data were to find their way into his first great geographical synthesis, *La Terre*, in the late 1860s.

The steamer *Philadelphia*, which had been built in Pennsylvania in 1854, made regular trips between New Orleans and Aspinwall (Colón) via Havana. On the particular trip that Reclus took, the ship broke down in Havana, so that he spent an unplanned fifteen days in Cuba, during which time he managed to see some of the island. He claimed that he was even pressed into service as a French–Spanish interpreter. There was no previous mention of any knowledge of Spanish, but perhaps he knew enough Spanish words to be of some help. His real knowledge of Spanish was acquired during his stay of about a year and a half in Colombia.[17]

The *Philadelphia* landed in Aspinwall with about three hundred passengers, and from there he took smaller vessels, first the *Narcisse* and then the *Sirio*, to ports along the Spanish Main—Cartagena, Savanilla, Barranquilla, and finally Santa Marta. He stayed for several weeks in Santa Marta but despaired of finding a suitable place to farm in the vicinity or sufficiently reliable farmers with whom he could enter into association. He described the lands behind Santa Marta as being of "exuberant fertility" and estimated that they "would be able to feed amply a half million persons," but "they were long ago conceded to a few great capitalists who wish neither to sell nor to cultivate them, and, in the vague hope of a future colonization undertaken on a gigantic scale, refuse to alienate the least part of their immense territory." For a few weeks Elisée lived with a young Italian, Andrea Giustoni, hoping to learn tropical farming. He was about to buy a one-hectare garden with a cabin and fruit trees for thirty-eight francs, but he declined because Andrea met with a severe injury, and Elisée could not find anyone else who could teach him how to farm. He then resolved to go to Rio Hacha, some 175 kilometers from Santa Marta, in the hope that Rio Hacha would offer better access to unclaimed lands in the Sierra Nevada de Santa Marta. In giving up the "Eden" of Santa Marta for prosaic Rio Hacha, Reclus felt that he was another Jérôme Paturot, the hero of Louis Reybaud's satirical novels who returned "to the prosperous but unexciting family business of selling cotton nightcaps" after some romantic adventures. During the two-day voyage to Rio Hacha in the schooner *Margarita* Elisée was chagrined to find that termites had eaten out the insides of all his books during his stay in Santa Marta.[18]

In Rio Hacha Elisée stayed at first with a Frenchman, Antonio Rameau. Rameau had come to Rio Hacha to dig artesian wells, but, having been unsuccessful in this venture, he became a blacksmith. The acknowledged leader of the small circle of Frenchmen in Rio Hacha was Jaime Chastaing (or Chassaigne), "a cabinet maker by trade and a remittance man by nature." After a few days with Rameau, Elisée rented a house and made fitful attempts to give lessons in French, English, and German while planning an agricultural enterprise in the mountains. He was disappointed in the lack of ambition of the young men of the town. Fifteen pupils showed up on the first day of class and none the second. Elisée declared that he would prefer to be a mountain peasant rather than a teacher. "Teaching is not my vocation," he said in a letter to his mother on 30 August 1856, "I prefer to sell bananas and arracachas than participles."

Reclus described Rio Hacha as a thriving town of more than five thousand population and accounting for perhaps two-thirds of the maritime commerce of New Granada. He felt that the town was "besieged" by the nearby Goajira Indians, who could easily destroy Rio Hacha but permitted it to exist because of their appetite for the trade goods they could get there. "If commerce ceased for any reason, the town would be burned the next day." He claimed that the Indians had destroyed all the Spanish settlements on the Goajira Peninsula, but he spoke of them with great admiration nevertheless. He was even befriended by a fat chief named Pedro Quinto and given the name "Felansi" (Frenchman), perhaps in memory of the French pirates who had aided the Goajira in burning Rio Hacha eleven times in the past.

After a year in New Granada, mostly in Rio Hacha, Elisée declared that

> The moment had come to realize my plans in some valley of the Sierra Nevada. Jaime Chastaing . . . always . . . in search of an El Dorado, was more and more unhappy with his lot; he begged me to accept him as an associate, and I was so weak as to give my consent. I naively thought that he had finally discovered his vocation at the advanced age of seventy and that all his dormant energy was really being revived. . . . I was going to live among the Arawak Indians, far from civilized society and have no other company than nature, my books and my projects. How sweet I thought my mother tongue, spoken by a compatriot in the midst of that solitude, would sound to my ears.

Elisée first made a reconnaissance trip into the mountains in the company of Chassaigne's son Luisito. After a long and difficult journey, part of which had to be made barefoot when they lost their sandals, Elisée found a mountain valley that pleased him very much. He picked out a fifty-hectare area of prairie half a league from the pueblo of San Antonio and then returned to Rio Hacha to make plans for setting up a farm. After resting for a month in Rio Hacha, he set out again for San Antonio, stopping first at the coastal village of Dibulla, where he hoped to hire Arawakan oxen as beasts of burden, but there were no Arawaks in Dibulla at that time, and, furthermore, Elisée came down with malaria. After a convalescence of two months Elisée again started out for San Antonio, this time on muleback and accompanied by the senior Chassaigne. Reclus's mule died, and he tried to continue on foot, but he collapsed from fatigue and was carried to San Antonio. He suffered a relapse of malaria and was confined to a cabin, but Chassaigne plunged into work, clearing, planting bananas, coffee bushes, sugar cane and vegetables, building a town house of granite blocks, erecting fences, and burning prairie. After a month Chassaigne became disillusioned, and he promptly dissolved the partnership and returned to Rio Hacha. Reclus did not feel that he could proceed with the agricultural scheme alone, and so he decided to return to France, even though he still had high hopes for the colonization possibilities of the Sierra Nevada de Santa Marta.[19]

For the rest of his life Elisée Reclus retained an idealized picture of Colombia, especially of the northeastern Sierra, and he never failed to recommend its agricultural possibilities. His enthusiasm has been echoed by writers down to the present day. Reclus was certain that the Sierra, which he estimated was about one-fourth the size of Switzerland, could easily support the same number of people as that European republic. José Miguel Rosales in 1934 was still extolling the great advantages of the Sierra for immigrants and said that 1,300,000 people could easily be accommodated there. Thomas Cabot in 1939 was also enthusiastic, but he cited as the greatest drawbacks the lack of markets and transport facilities and especially "the lack of native enterprise." The Sierra Nevada de Santa Marta has yet to bear out Elisée Reclus's hopes.[20]

Reclus left Rio Hacha, never to return, 1 July 1857, on board the *Providence* bound for le Havre. His *Wanderjahre* over, he was ready to return to France, which was then prospering under Napoleon III, whose coup d'état had forced Elisée to go abroad in the first place. Elisée was eager to rejoin Elie, who had married their first cousin, Noémi Reclus, in 1855 and was living in Paris. The brothers remained almost inseparable for more than a dozen years, down to the *débâcle* of 1870.

Chapter Three

The Paris Years

Although Elisée joined Elie in Paris in 1857 and was to live there for the next fourteen years, the life of this abstemious young republican was not exactly the Offenbachian *vie parisienne*. The revels and debauches of the Second Empire were not suited to his puritan tastes, but the great city afforded him the opportunity to practice and perfect his writing skills and to continue his political education.

Like his brother before him, Elisée's first thoughts on his return to France were to find a wife and a job. In both quests he was successful, but not immediately. Choosing a mate was a rather uncomplicated procedure in those days; it usually involved making a choice from a relatively small range of prospects in one's family and social circle; physical attraction was not a primary consideration. Fortunately for Elisée, his lack of wealth obviated an arranged marriage, and perhaps a lack of marriageable cousins prevented him from imitating his older brother. He was able to indulge in the rare privilege of romantic love. Elie's wife Noémi introduced Elisée to a beautiful young woman in Sainte-Foy, Clarisse Brian, whom he is said to have "glimpsed" during his college days a decade earlier. Clarisse had been born in St. Louis, Sénégal, in 1832, the daughter of a Bordeaux merchant and his Senegalese (Fulani) wife, and at the age of eight she went to France, first to Bordeaux and then to Sainte-Foy, to live with her paternal grandmother, who was an American. Clarisse had been baptized a Roman Catholic, but she became a Protestant after moving to France, and after her marriage she gradually gave up all religious practices because of the influence of her husband.

[37]

It is said that Elisée was attracted to Clarisse because of her African ancestry. Indeed, Paul Reclus, Elisée's nephew, has said: "There is not the slightest doubt that his sojourn in Louisiana formed in him the decision to marry a daughter of a spurned race." I can find no evidence that Elisée had made such a decision before he returned to France. I think, rather, that the correct explanation is the simplest one: he fell in love with Clarisse; her ancestry was probably not an important factor. In any event, Elisée courted her through most of 1858 during visits to Sainte-Foy, and, two days after they were married on the fourteenth of December, they moved to Paris, where he had been promised a small office in the Hachette publishing house.[1]

The matter of employment had been a problem to Elisée ever since he had disembarked in August 1857. His mother—as mothers are wont to be—was particularly concerned that he should find a job, and she even relayed to him the repugnant offer from his materialistic uncle Chaucherie to join in a scheme involving the cutting of an area of woodland in the Périgord. Elisée wanted to remain in Paris, perhaps more to be with Elie than because of the amenities of the great city. The extreme fraternal devotion was plainly stated in a letter to his mother soon after his return from the New World:

> . . . my wish—my most ardent wish—has always been to live with Elie; it was especially to prepare for our comfort and liberty that I left for America, and, when I recognized the impossibility of finding that retreat because of climate, lack of resources, and sickness, I came back. Now that I am with my brother, it would be bitterly sad for me not to remain with him, and I will never attempt any undertaking if he doesn't figure in it from the very beginning.[2]

It is truly amazing how Elisée was to maintain propinquity with Elie for the rest of their lives. Indeed, they are united even in death, for they occupy the same grave in Ixelles (Brussels), apart from their wives.

In that same letter to his mother, Elisée stated his desire to become either a professor of geography (presumably at the lycée level) or a journalist (not a newspaperman but a magazine writer).

> It is true that I hardly like the profession of teaching when it is a matter of teaching absurd alphabets and jargons against which my intellec-

tual sense rebels, but I am happy when I talk about geology, history, and the truly useful sciences; the idea that perhaps I could become a professor of geography fills me with joy. I have spoken to you also of journalism; there are journals after journals; there is the *Moniteur*, *Patrie*, and other mercenary sheets; but there is also the *Journal de géographie* [sic—the *Bulletin* of the Société de géographie de Paris], the *Journal Asiatique*, and the *Journal Statistique* [sic—*Journal des économistes*], and my pride would not suffer at all in having to sign my name to articles on the Mississippi or the Sierra Nevada [de Santa Marta]. It is precisely to have access to such journals that I have gone to present myself to Messrs. d'Orbigny, Cordier, and Alfred Maury; but unfortunately these gentlemen were away. If . . . I could enter into the editing of a political journal, I would not be humiliated; I would quite simply have dangers to risk, since it is dangerous to tell the truth. . . . The job of clerk would be for me only a last resort, but I would accept this with joy because it would permit me to remain with my brother and to find myself again in an atmosphere of art, science, and of life which was lacking to me for such long years.

But he went on to say that he would do any kind of work, so long as it was *useful* work, citing his previous experience as a manual laborer, carpenter, and fish merchant.[3]

Reclus was hopeful that he could get a position as a correspondent on European affairs for the New Orleans newspaper, *L'Union*, which had earlier published two articles which he had sent them from Colombia. He proposed that he report to *L'Union* on European political affairs for a salary of two hundred francs a month, but unfortunately the newspaper did not survive after October 1857, and so none of his European articles was published. Exploiting his Sainte-Foy connections, Elisée tried to get a teaching position in a lycée in Paris through Elie Broca, but classes had already begun, and he was unwilling to take an inferior job as a *répétiteur* (assistant). He then accepted a tutoring job in the home of another Sainte-Foy friend, Fezandié. He was particularly pleased to be able to give one special lesson in geography each week. At the same time, he worked on his geographical writings. He hoped that he would be able to read a paper before the prestigious Société de géographie de Paris, and he was anxious to amass the necessary sixty francs so that he might join the Society. Messrs. Maury and Lejean proposed his candidacy on 2 July 1858, and he was admitted to membership a fortnight later. Reclus was to take a very active part in the affairs of the Society for the next twelve and a half years, even through the siege of Paris by the Germans in 1870-1871.[4]

Casting about for other writing commissions, Elie and Elisée wrote a remarkable joint letter to the editor of the *Revue germanique* on 6 January 1858. Elie offered to write about German philosophers, and Elisée indicated an interest in reviewing German scientific works, especially in the fields of geography and geology. The letter is even more interesting for what it reveals about the religious evolution of the brothers.

> We are two brothers who, having sojourned in Germany at different times and for several years, can say, without boasting, that the German language no longer has any difficulties for us. Both of us studied Protestant theology from 1846 to 1851 at Geneva, Berlin, Montauban, and Strasbourg; but, for reasons that freethinkers will easily be able to appreciate, we have never practiced the ministry. Philosophically, we attach ourselves to the school of Spinoza.[5]

Under his own name and under the pseudonym of Jacques Lefrêne, Elie contributed a wide variety of articles to the *Revue germanique,* but Elisée produced little apart from a translation of a paper by Carl Ritter, "De la configuration des continents sur la surface du globe, et de leurs fonctions dans l'histoire."[6]. The paper had been read by Ritter before the Berlin Academy of Science on the occasion of the two-hundredth anniversary (1846) of the birth of Leibniz, was published in German in 1850, and then was translated into French "on his [Ritter's] request and under his eyes" by Elisée Reclus. It was printed in the *Revue germanique* in late 1859 just after Ritter's death and contains a perceptive sketch of Ritter written by Reclus, in which he recounts the enthusiasm of the great professor in 1851 and the rapt attention given him by Reclus and other admiring students. The statement by Reclus that the translation was done at Ritter's request and under his direction might indicate that the two were in communication at some time between 1851 and Ritter's death in September 1859, and it is entirely possible that Reclus could have paid his aged mentor a visit when the young French geographer toured northern Germany in the summer of 1859, but there is no record of such a visit.

[40]

More important than his brief association with the *Revue germanique* was Reclus's connection with the *Revue des Deux Mondes,* which has been called "the vehicle par excellence of German influence in France in the nineteenth century."[7] However, Reclus used the *RDM* not so much for reviewing and explicating German works as he did for the publication of more general geographical articles and reviews, especially concerning the Americas. For almost a decade, from 1859 to 1868, Elisée Reclus published a large number of articles in the *RDM* which collectively add up to an impressive body of geographical literature. It was here that he published a series of four articles, "La Nouvelle-Grenade, paysages de la nature tropicale," in 1859-1860, which were then reprinted in book form by Hachette in 1861 as *Voyage à la Sierra-Nevada de Sainte-Marthe: Paysages de la nature tropicale.* This was Elisée Reclus's second book; the first, *Guide du voyageur à Londres et aux environs,* had been published in 1860 by Hachette as part of the "Collection des Guides-Joanne."

Reclus's association with Adolphe Joanne, the French counterpart of Karl Baedeker and John Murray, and with the Hachette publishing house, a lifelong association that began in 1858, was the most important influence in shaping his career as a geographer. The Hachette connection was also of great importance to Elie and Onésime, who derived from it probably the major part of their income throughout long periods of their lives. Their cousin Franz Schrader was also drawn into the Hachette orb. Louis Hachette (1800-1864) had founded the publishing house in 1826. His son-in-law and associate Emile Templier (1821-1891) was especially concerned with geographical works. Adolphe Joanne (1813-1881) had traveled since 1833, and in 1841 he edited a guidebook to Switzerland. In 1855 Hachette took over the Joanne guides, which had formerly been published by Louis Maison, and Adolphe Joanne became the general editor of the guidebook series. His son Paul carried on the enterprise for some three decades after his death.[8]

Adolphe Joanne took young Elisée under his wing and made him one of the mainstays in his stable of guidebook authors. He subsidized much of Reclus's travel in the late 1850s and in the 1860s, and, indeed, even traveled with him on at least one occasion, in 1859. Reclus wrote or contributed to several Joanne guides in the 1860s, beginning with the 1860 London guide, which was revised in 1862, and in 1864 he contributed a long introduction (about 150 pages) to Joanne's *Dictionnaire des communes de France*. With Elie's help he revised this introduction in 1869, and in 1905, in the last few weeks of Elisée's life, Paul Joanne published Reclus's introductory volume to the latest *Dictionnaire*, a seven-volume work which commenced in 1890 and of which the introduction was actually the last volume to be published. The elder Joanne was very fond of Elisée at first and praised him as "an intrepid traveler as well as a learned geographer," but he was unsympathetic to him in 1871.[9]

The Joanne guidebooks, like those of Baedeker and Murray, were noteworthy for the sobriety and spareness of their prose, and a biographer looking for early signs of Reclus's development as a geographer is bound to be disappointed. Humor is lacking altogether, except in rare instances such as Adolphe Joanne's preface to the third edition (1859) of the *Itinéraire descriptif et historique de la Suisse*, in which he begs the Swiss innkeepers to be more polite to the weary travelers, who are too often rudely turned away or given the smallest, dingiest, least airy, and gloomiest room on the top floor at the most exorbitant price.[10]

It was in his articles in the *Revue des Deux Mondes* and in the geographical journals that Reclus was able to bring in his personal observations and first-hand accounts of places he had visited in Louisiana, Colombia, and Europe. His travels in Europe, sometimes on his own but usually on commission from Joanne, are described in articles in the *RDM*, *Tour du Monde*, *Bulletin de la Société de géographie*, and the *Annales des voyages*. His observations on the southwestern coast of France, including Les Landes, are especially interesting because they show how the young autodidact was learning to become an intelligent observer of physical geographical phenomena, as well as of human activity. Physical geography had displayed rather meager development before Reclus's time, but he was able, through his own observations and especially the writings of contemporaries, to achieve a fairly high level of physical description and explanation. In his discussion of the effects of man upon nature, he stressed not only the destructive effects, like his American contemporary George Perkins Marsh, but also the constructive efforts. Like his Christian mentor, Carl Ritter, Reclus thought of the earth as the home of man, a place created for man's use and enjoyment. Instead of always desecrating the earth, man often improves it. Recent advances in science and engineering could, Reclus thought, be of enormous benefit to the land and the people living on it. He was to retain this basic faith in the positive benefits of science and technology throughout his life. Similarly, he could see the long-range benefits of modernization, as in the "conquest" of Les Landes, the sandy pinelands of southwestern France, where now (1863)

> the majority of children go to school; the newspaper and even books have penetrated the forest; and the doctor has replaced the sorcerer in the treatment of illnesses. The French territory is enriched with a new province, which doubtless will be one of the most charming . . . Having been pacified without bloodshed, this conquest of Les Landes will be no less useful and will be more durable than that of many distant colonies which were bought at the price of thousands of precious lives.[11]

A particularly fine example of Reclus's views on landscape appreciation is his article, "Du Sentiment de la nature dans les sociétés modernes," published in the Revue des Deux Mondes in 1866. In it he discusses the growing cult of Alpinism among Europeans, particularly the British, who do not have impressively high peaks at home but who are ardently devoted to physical fitness and to overseas travel and commerce, all of which contribute to their leadership in mountain climbing and related activities. Reclus credits the Germans with having more intimate feelings and knowledge of landscape than the English, not only in actual scientific exploration but also in literary exploration, as in the works of Kant, Goethe, and Ritter, "that heroic scholar who did not recoil from the thought of commencing all by himself the encyclopedia of all human knowledge on the lands and peoples of the world."

> It must be said that the French, taken as a whole, do not always understand the splendors of Nature as well as their neighbors to the north and east. More sociable than the Germans and the English, they much less easily endure solitude or even the temporary interruption of their habitual relations. . . . The Nature which the French understand best and which they most like to look at is the gently undulating countryside where diverse crops alternate gracefully right

up to the far horizon of the plains. . . . [Thus] Nature, fashioned by human work, is humanized, so to speak.

In a manner reminiscent of Carl Ritter, Reclus stressed the mystical bond between man and nature. "The developments of mankind are tied in the most intimate fashion with the natural environment. A secret harmony is established between the earth and the peoples which it nourishes, and when imprudent societies permit themselves to lay a hand on what makes the beauty of their domain, they always end up repenting it."[12]

It is this peculiarly French view, this appreciation of the humanized earth, that sets Reclus apart from his American contemporary, George Perkins March, who in his great book *Man and Nature* (1864)—a book of astonishing durability which continues to attract great interest in the 1970s—stressed the *destructive* effects of man's occupance more than his constructive effects. In a review of Marsh's work in the 15 December 1864 issue of the *Revue des Deux Mondes*, Reclus praised it highly, although he thought it "too devoid of method." He was able to offer a mild criticism of Marsh's description of the dunes of Gascony from his own experience. While not denying the destructive effects of man's occupance, Reclus also emphasized the beneficial effects—or at least the ability to improve the land.

> The action of man gives the greatest diversity of aspect to the earth's surface. On the one hand it destroys, on the other it improves; according to the social state and the progress of each society, it contributes sometimes to degrade nature, sometimes to embellish it. Camped like a trader in passage, the barbarian pillages the earth; he exploits it with violence . . . He finishes by devastating the land which serves him as a home by rendering it uninhabitable. The truly civilized man, understanding that his own interest is tied to the interest of all and to nature itself, acts completely differently. Not only does he know how, in his role as an agriculturist or industrialist, to utilize more and more the products and forces of the globe; he learns also, as an artist, to give to the landscapes which surround him more charm, grace, and majesty. Having become 'the conscience of the earth,' the man worthy of his mission assumes part of the responsibility in the harmony and beauty of the environing nature.[13]

Reclus cited numerous works of man which have been decided improvements—for example, diking and draining in the Netherlands, drainage works in French Flanders, dune stabilization in Gascony, and reforestation in the British Isles.

> To all these great works, having as their goal modifying the surface of our earth to the benefit of man, is tied intimately a work which can seem chimerical to many but which is no less of the highest importance. It concerns conserving or even enlarging the external beauty of nature, of returning that beauty when brutal exploitation has caused it to disappear. In various parts of Europe and notably in France, one can traverse certain plateaus for hours without finding a site where the glance of an artist could rest with satisfaction . . . However, it is so easy to put the soil under cultivation while leaving the landscape in its majestic beauty! In England, that land where the farmers know how to make their fields produce such abundant crops, but where the people have always had more respect for trees than the Latin nations have had, there are few sites which do not have a certain grace, or even true beauty. . . . But the people who are in the vanguard of humanity today are in general very little concerned with the embellishment of nature. More industrialists than artists, they prefer force to beauty.[14]

I shall turn later to the interesting exchange of correspondence between Marsh and Reclus in the years 1868-1870, but here I want to stress that they were on parallel courses, appreciative of each other's work but not borrowing their essential ideas. Reclus's work, at least at this time, was closer to Ritter's and to that of another pupil of Ritter, Arnold Guyot, whose great book, *Earth and Man* (1849), was much admired by Reclus.

In 1863 Elisée Reclus encountered the enterprising American consul-general in Paris, John Bigelow, whose personal diplomacy on behalf of the Union cause had earned him a mild rebuke from Washington but gained some sympathetic friends in Paris. With support from Reclus, Bigelow got himself elected to the Société de géographie de Paris in April 1863 in order "to have some means of showing how geographical science has flourished in America during and in consequence of this war [American Civil War] and also put the members in as good humor as possible with our government by presenting them with some scientific memorials of our military doings . . . My opportunities of getting into relation with men of influence about the Government are sufficiently limited at the best and any footing of this kind that I can obtain greatly increases my means of usefulness." He found a friend in Reclus, "so competent a man and so enlightened and thoroughgoing friend of our cause." Bigelow tried to acquire American publications, especially government documents, for Reclus to review in the *Revue des Deux Mondes*. Bigelow had corresponded with George Perkins Marsh in 1863, and it was perhaps he who introduced Reclus to Marsh's book the following year.

Although Bigelow never saw Reclus after he left the embassy in 1866, he kept track of his friend and tried to influence his successor as minister to France, Elihu Washburne, to intercede for Reclus in 1871. Bigelow's concern, as well as that of William Huntington, Paris correspondent of the New York *Tribune*, might have aided Reclus's cause somewhat during his imprisonment after the Commune. When Reclus settled in Lugano, Switzerland, in 1872, Bigelow sent him, from Germany, a gift of 125 thalers, which Reclus accepted, not for himself, but to give to needy friends. This gift was misconstrued by later writers who claimed that Elisée Reclus turned down a huge cash award from the United States government for his propaganda on behalf of the Union cause. After Reclus's death and close to his own, the aged Bigelow described Reclus in his memoirs as "a man of high character and extraordinary intellectual activity . . . the most violent reactionary against dynastic government that I had ever met."[15]

In April 1865, Bigelow, who had just been promoted to the position of Minister to France, received a letter from Reclus offering condolences on the death of Abraham Lincoln. Reclus was then in Sicily to observe the eruption of Mt. Etna and to make other observations for a revised Joanne guidebook to southern Italy and Sicily. He had just heard of General Lee's surrender ending the American Civil War and also of the assassinations of President Lincoln and Secretary of State Seward (this last rumor was, of course, erroneous). "The history of the world now pivots on the United States," said Reclus in a statement which was perhaps premature in 1865. Not so prescient was his prediction that the Mafia, "that corporation of thievery and fraud, will [soon] have disappeared like so many other institutions bequeathed by the Middle Ages."[16]

The journey to Sicily provided Reclus with a rare opportunity to see a major volcanic eruption. Beginning in late January 1865, Etna erupted for the next two hundred days, but Reclus was able to walk right up to the very crater. "Leaving all my useless baggage at the inn . . . I buckled my knapsack on my back and, joyful as a student from Germany, I commenced my walking journey . . . As a free man I exposed myself to an excess of curiosity on the part of the gendarmes." Reclus's observations of Mt. Etna were more those of a German student than a German professor: he did not make "scientific" observations but waxed enthusiastic about the magnificent view from the top. The walk was more a physical and emotional experience for him than a cerebral one. His life-long interest in volcanoes was rekindled after the eruption of Mt. Pelée in Martinique in 1902. At that time he initiated one of his last major projects, a world-wide survey of volcanoes which was published posthumously in the years 1906-1910.[17]

The trip to Italy in 1865 was also significant in Reclus's political development, for he visited Michael Bakunin in Florence and was admitted to Bakunin's secret brotherhood. They had previously met in Paris in November 1864. Reclus subsequently followed Bakunin into the League for Peace and Freedom, Bakunin's own International Alliance for Socialist Democracy, and finally into the International Workingmen's Association. Although it has been claimed that Reclus "was the most complete, already full grown anarchist whom Bakunin met, Proudhon excepted" and therefore was not profoundly affected by Bakunin's personal influence, it would appear that this influence was of vital significance in Reclus's political evolution. Contact with "the most dramatic and perhaps the greatest of those vanished aurochs of the political past, the romantic revolutionaries," as George Woodcock has described Bakunin, was an important step in the radicalization of Elisée Reclus, a process which was completed in the 1870s, after the trauma of the Paris Commune. (In a later chapter, I shall return to Bakunin, especially to his perceptive description of the Reclus brothers in 1871.) Typically, Elie was not so taken with the Russian revolutionary as his younger brother was. When Elisée embraced a new philosophy, he entered it fully, heart and mind.[18]

The Ritterian general geography which Elisée Reclus had conceived in Ireland in 1852 finally came to fruition late in 1867, when he finished the first volume of *La Terre: description des phénomènes de la vie du globe*. He had signed a contract with Hachette for this work on 24 December 1862, and although the contract stipulated that it was to be completed by the end of 1865 at the latest, the first volume was actually finished in 1867 and the second a year later. The two volumes bear the publication dates of 1868 and 1869, respectively, because, like so many of the popular works published by Hachette, they came out at the very end of the previous years and were intended as "étrennes"—*i. e.*, Christmas or New Year's gifts.[19]

The two volumes of *La Terre* are roughly the same size—775 and 742 pages, respectively—and they cover in a systematic fashion the physical geography of the earth. The first volume, "Les Continents," consists of four sections, one on the earth as a planet and on geological time, the second on the major landforms, the third on the circulation of waters, and the last on volcanoes, earthquakes, and the slower oscillations of the crust. The second volume, "L'Océan—L'Atmosphère—La Vie," has three parts, the first dealing with the oceans, including islands and shorelines (with a separate chapter on sand dunes, reflecting Reclus's special interest), the second treating climates, weather elements, and also magnetism and auroras, and the last section covering flora and fauna, including man. The very last chapters of volume two, "The Earth and Man" and "The Work of Man," are, to my mind, the most interesting parts of *La Terre*; certainly they are the most enduring. The former treats the influence of nature upon man, and the latter concerns the reaction of man to nature, the diffusion of culture over the earth, and even the influence of man on the beauty of the earth—the embellishment and uglification of the world. He concluded the work with a sermon calling for cooperation and moral regeneration.

> Science . . . shows us the means to embellish the earth's surface and to make of it the garden dreamed of by the poets of all the ages . . . [but] it alone cannot finish the great work. To progress in knowledge must correspond moral progress . . . The traits of the planet will not have their complete harmony if men are not first united in a concert of justice and peace. To become truly beautiful, the 'beneficent mother' must wait until her sons embrace each other as brothers and until they have finally concluded the great federation of free peoples.[20]

La Terre was a great success, earning virtually unanimous praise in the newspapers and scientific journals, "It is written with an elegance and a lucidity of style which lend a great charm to its reading," said V. A. Malte-Brun, son of Conrad Malte-Brun (1775-1826), the author of *Précis de géographie universelle* (8 vols., 1810-1826), which was to be a model for Reclus's "*New* Universal Geography." Also noting its "ease and elegance of style," Charles Grad said that *La Terre*, "if not a definitive work, must nevertheless be considered as a sure guide . . . and worthy of figuring as one of the monuments of science alongside the *Précis de géographie universelle* of Malte-Brun, Humboldt's *Cosmos*, and Ritter's *Erdkunde*." In the new Comtian journal *La Philosophie positive* G. Wyrouboff lauded *La Terre* for both its scientific and its literary merit, regretting only that it had not been entitled *Geology*, which would have made it more acceptable to the positivists. An anonymous English reviewer considered it "a well-digested compilation" but thought Reclus's heavy indebtedness to the works of Marsh, Maury, and Mallet was not sufficiently acknowledged. In my opinion this criticism was quite unfair; Reclus had certainly read these works, but he made proper mention of them when he used them, and he did not use them excessively or slavishly. He reacted strongly against the English review in a letter to Marsh on 14 April 1868.[21]

One of the most appreciative reviewers of *La Terre* was the German geographer Oscar Peschel, who published numerous excerpts and critical comments in his journal *Das Ausland*. Although he criticized Reclus for erroneous statements or for repeating the mistaken views of Ritter and others, Peschel praised the work generally and remarked on its typically French clarity of exposition. Although Peschel had been "occupied daily for about 20 years with the same tasks," he felt that he had learned a great deal from reading *La Terre*.

> We can . . . recommend this work with a good conscience to our German colleagues, even though it comes from France, where the enthusiasm for geography, formerly so important, has noticeably cooled in this century . . . Moreover, most of the material is well-founded, and, that is the mastery of the French, who are much aided by their language—clear and precisely presented, and moreover the comprehension of each passage is facilitated through several fortunately chosen illustrations. We might add that M. Reclus totally

masters German and English and is well informed of the newest research in England as well as with us.

Peschel criticized Reclus's ideas on the origin of Venice's lagoons, but he liked his descriptions of fiords and sand dunes. Indeed, he thought that Reclus's treatment of fiords was better than his own in *Neue Probleme der vergleichenden Erdkunde* (1870), a work that established Peschel as the foremost physical geographer of his day. It is interesting that Robert E. Dickinson has singled out the section on fiords in *Neue Probleme* as "an excellent illustration of Peschel's method." The two works, *La Terre* and *Neue Probleme*, were credited with awakening popular interest in geography and with laying the foundation of the New Geography that was to emerge in French and German universities after the Franco–Prussian War. Alfred Hettner, in a general review of nineteenth-century geography in 1898, credited these two works with putting geography back on the scientific track after the hiatus following Humboldt's death in 1859.[22]

In a letter to Peschel dated 21 October 1868, Reclus acknowledged the German geographer's useful criticism of the first volume of *La Terre* and announced that he had a much grander project in mind.

> For a long time I had a more considerable project, that of writing a *General Geography*. That work, as you know, is lacking in France, for the *Précis de Géographie* of Malte-Brun, excellent for the time when it was written, has aged remarkably in the last fifty years, and the revised editions which have been published recently leave much to be desired from the point of view of method, accuracy, and breadth of ideas.
>
> The *Geography* that I intend to write, and for which I have amassed the material, would comprise about ten volumes, each of the size of your *Geschichte der Erdkunde*. It would not be so large as several of the great encyclopedias published in Germany; but it would be larger than Malte-Brun's work and would be certainly long enough to give in a clear and simple style a complete description of the various countries of the earth, without falling into the infinity of details and into the tedium of nomenclature. Furthermore, what would give my work a distinctive character and would assure it a place completely apart from the works of the same sort which have appeared up to now in France and Germany is that each chapter would include a certain number of special maps, plans, pictures, profiles, and cross-sections. The entire

work would contain two or three thousand of these maps or illustrations.

Reclus hoped to enlist the support of English and German publishers, and he proposed that Peschel revise his *Geschichte der Erdkunde*
so that it might serve as the introductory volume of the *Géographie
générale*. Reclus or one of his friends would translate the *Geschichte*
from German into French. Peschel's reaction to this proposal is not
known; indeed, it may be that Reclus's letter was only a draft and was
not actually sent to Peschel. In any event, the letter is extremely
interesting because it shows how Reclus's plan for his *Nouvelle géographie universelle* was taking shape as early as 1868.[23]

In *Das Ausland* Peschel noted the slight emphasis given to human
geography in *La Terre* but said that the author was probably saving
much material for another book. His own faith in the ingenuity of man
ran counter to the rather definite environmentalism that he detected
in Reclus's work. Peschel also noted Reclus's emphasis on the beneficial effects of the man-nature relationship, as opposed to Marsh's stress
on the deleterious effects.[24]

Early in 1868 Marsh wrote Reclus a congratulatory letter after
reading volume one of *La Terre*, and there followed a very fruitful
correspondence for the next two years until the outbreak of the
Franco–Prussian War. The two men never met, although Marsh was
then living in Florence and had hoped to meet Reclus on one of his
visits to Paris. Marsh warmly lauded *La Terre* and urged an English
translation. He encouraged Reclus to think that Charles Scribner's
Sons of New York would be willing to publish the translation, but it
did not appear until 1871-1873, and Harper was the American publisher. Marsh wrote an introduction to the English translation, but his
essay was not published until 1960, when it appeared in abridged form
in the *Geographical Journal*. Similarly, Reclus strongly supported a
French translation of Marsh's *Man and Nature*, and he was actually
working on the translation with another Frenchman named Mackintosh in 1870. Mackintosh was killed in the Franco–Prussian War, and,
although Reclus preserved part of Mackintosh's translation and was
still hoping to publish it as late as 1900, the French translation of *Man
and Nature* has never appeared. In 1884 Marsh's widow offered to
return Reclus's letters to her late husband, and she asked if she might

borrow the letters which Marsh had written to Reclus, but in his reply Reclus said that the letters had been "plundered in 1870 or 1871, during the civil war." Presumably this would account for the dearth of incoming letters in the various Reclus archives before his removal to Switzerland in 1872.[25]

In 1869 Reclus published a charming little book of nature sketches, Histoire d'un ruisseau ("The Story of a Brook"), which he later declared to be his favorite of all the books he ever wrote. It was a great favorite with the reading public, too, and along with its companion volume, Histoire d'une montagne (1880), was often selected as a prize book to be given to school children. In this book, Reclus not only stressed the moral value of nature contemplation, especially in the upper reaches of streams, where they are the most pure, but he used the stream as a metaphor of life. "The story of a brook," he began, "is the story of the infinite." He closed by saying that the hydrological cycle is the image of all life, the symbol of immortality. Each person is like an ever-changing river, made up of innumerable molecules on an endless voyage.

> Peoples mix with other peoples like brooks with other brooks, rivers with rivers; sooner or later they will form a single nation, just as all the waters of the same basin finish by merging into a single river . . . Humanity, until now divided into distinct currents, will be no more than a single river, and, reunited into a single flow, we will descend together toward the great sea where all life will lose itself and be renovated.[26]

It was apparently from this book, as well as from suggestions in Reclus's other works, that Patrick Geddes derived his famous "Valley Section," the graphic representation of the characteristic distribution of settlement and economic activities. But the idea of river basins as the fundamental geographical regions goes back long before Reclus, as undoubtedly does his observation that the whole panorama of human history unfolds as one descends a river to the sea. This is an example of Ritterian rhetoric, with not a little Ritterian religiosity, but Reclus has substituted pagan deities for Ritter's One God. With characteristic fervor Reclus proclaimed that if one loves a stream, it is not enough simply to look at it, study it, or walk along its banks, but one must gain a more intimate knowledge of it by plunging into its waters. "One becomes a triton again, just as our ancestors were." This statement had an odd echo almost three decades later in Geddes's description of Reclus bathing (presumably in a public pool) in Bristol, England, while attending the annual meetings of the British Association for the Advancement of Science in 1898: "Old R. like a river god: Imagine his dripping beard & hair & mixture of grim & gentle aspects!" Reclus had fond memories of the streams he had known, perhaps even of the Dronne River of his infancy in La Roche-Chalais, and he was deeply moved by the springs he had seen in Sicily in 1865. He remembered an incident from his youth when he saw a group of stern-faced soldiers take off their uniforms and leap joyfully into a river. The lesson that hostility diminishes when people remove the trappings of status stayed with Reclus to the end of his life.[27]

Reclus composed part of *Histoire d'un ruisseau* at Vascoeuil, the country estate of Alfred Dumesnil in eastern Normandy. Dumesnil, son-in-law of Jules Michelet, had been widowed in 1854 and was trying to raise his two young daughters at Vascoeuil. He met the Reclus brothers in Paris, and they and their families were invited to Vascoeuil for extended periods beginning in 1863. Louise Reclus, sister of Elisée and Elie, was introduced to Alfred Dumesnil when she visited her brothers on her way back to Orthez from Ireland, and she was then engaged as a tutor to the Dumesnil children. She ultimately married Alfred Dumesnil in 1871. Vascoeuil became a sort of salon, rather like Brook Farm, for Dumesnil and his friends, including the Reclus and the men of the neighborhood. The Reclus were allowed to bring their own friends to Vascoeuil, such as the young American medical student Mary Putnam, and it is from her that we have some of the most vivid sketches of life at Vascoeuil.[28]

Mary Corinna ("Minnie") Putnam, the first woman admitted to the Ecole de Médecine in Paris, was the daughter of a New York publisher, George Putnam. She met the Reclus brothers in Paris in March 1868 and was soon invited to Vascoeuil, where her first visit was clouded by the death and funeral of Alfred Dumesnil's ninety-two-year-old father. Mary thereafter made several extended visits to Vascoeuil, not holidays entirely, because everyone was expected to do a little work and to contribute towards the expenses, but there were long periods devoted to reading, nature study, games, and family activities. In a letter to her mother from Vascoeuil on 22 September 1868 Mary described the daily round:

> I shall stay here till the 15th of October. . . . It is perfectly delightful here. The country is charming, the time of year delicious, and the family extremely interesting and lovable. I derive infinitely more pleasure, benefit and solid satisfaction by my sojourn here than from a rapid "European tour." Travelling—rapidly at least—is a horrid dissipation,—and I have a horror of dissipation.
>
> . . . whoever stays here, pays his division of the monthly expenses (not his *board*, properly speaking, as no profit is expected) so that one is infinitely more at liberty than if on a visit. At the same time, it is the custom for every one to contribute what he can of personal value, one teaches the children geography (there are commonly two or three families) and another a peculiar kind of drawing, another fencing, etc. I myself made some American gingerbread the other day, and give lessons in calisthenics—besides discovering an extremely noisy way of hopping up stairs, with which the juvenile population is enchanted.[29]

On her return to Paris after seven weeks at Vascoeuil Mary again wrote to her mother (29 October 1868):

> It *was* delightful there. I used to get up as soon as it was light, then write till eight o'clock; then we breakfasted, after which I wrote till the dinner at noon, then read all the afternoon, at first by a delicious little brook, afterwards in a curious round tower that forms part of the château. In the evening we continued the conversations already commenced at breakfast, dinner and supper and which were always gay and animated. Every few days we took a long walk through the most charming country. I was extremely happy all the time, and I lived on fifty cents a day all told.[30]

Mary Putnam had come to Paris in the autumn of 1866 with the intention of enrolling at the Ecole de Médecine, a male bastion. She was allowed to attend a course in January 1868, and she was graduated in July 1871. Her persistence enabled an Englishwoman, Elizabeth Garrett, to gain admission to the School, and Miss Garrett finished her work earlier than Miss Putnam, so that the latter was the second woman to graduate from the School, even though she had been the first to enter. Mary Putnam became very close to the Reclus; indeed, she moved in with them in November 1869 and stayed until the bombardment of Paris by the Germans in the winter of 1870-1871. Miss Garrett also stayed with the Reclus for a week in late 1869. The household, a fifth-floor apartment on the Rue des Feuillantines (now Rue Claude Bernard), must have seemed like a veritable commune or a Parisian version of Vascoeuil.[31]

Mary Putnam, whose temperament was so much like that of the Reclus, became a real member of the family and a champion of its causes. She adored the oldest brothers and described them in much the same way that other observers have done. Elie, she said, "is very interesting,—a calm, reticent, benign kind of man, but one of strong, deep enthusiasm such as you rarely see in a Frenchman, a man who glows with the subject he talks about, but never flames." "I do not think I ever met a more perfect character than that of Elisée Reclus. . . ."

> Elisée Reclus reminds me . . . of what Jesus Christ must have been, both from character, and from his position of antagonism, and isolation from the world which surrounds him, and upon which he is too unworldly and impractical to cast any influence. . . . I do not know how many people I have heard say, "It is impossible to have a mean thought in the presence of Elisée Reclus."

"The whole family," said Mary Putnam, "have a singular vigor and finesse of intellect . . . and at the same time they are so devotedly attached to each other, so warm and demonstrative, as to be an astonishing revelation in the midst of cool, polite Paris."[32]

Mary Putnam was also present in the last days of Clarisse Reclus. Elisée and Clarisse had two daughters, Marguerite ("Magali"), born 12 June 1860, and Jeanne ("Jeannie"), born 1 March 1863. A third daughter, Anna, died soon after birth, and the mother then quickly succumbed to galloping consumption on 20 February 1869. This was the first great tragedy of Elisée's life. He had been completely devoted to Clarisse, and she to him. Theirs was a perfect marriage, although Clarisse did not get deeply involved intellectually in the political and social concerns of her husband. Her full-time rôle was that of loving wife and mother. "Of an imposing beatuy, a majestic gait, a limitless kindness, she made a delicious mother to her two children," as nephew

Paul Reclus recorded in his memoirs. After the death of their mother, Magali and Jeannie were sent off to live with relatives for a while, the former to Orthez and the latter to Nîmes.[33]

When Mary's father, George Putnam, the publisher, visited Paris in March 1869, she took him to dinner at the Reclus, who, he reported in a letter to his wife, were "queer but *nice* in their way." In Mary's own report of that evening, the Reclus were "very *empressés* and cordial." The mother was at first rather anxious about her daughter's associations with people of such advanced political opinions, but Mary vastly preferred their company to that of bourgeois Parisians. "In regard to the whole family I may say, that to anyone who desires bright, fresh, inexhaustible intellects, hearts pure and warm,—the most delicate honour, dignity and grace of character,—ideas romantic and living,—I cannot readily imagine them better suited elsewhere."[34]

Mary Putnam wrote some pieces for her father's journal, *Putnam's Magazine*, and she persuaded Elie Reclus to publish several articles there. In November 1869 Mary was commissioned by her father to cover the opening of the Suez Canal, but she passed the invitation on to Elie, and he wrote two accounts of his journey, one for *Putnam's Magazine* and a longer and more interesting one for *La Philosophie positive*. The assignment might have gone to Elisée, but he was then spending several months in Nice, working on a Joanne guide to the Riviera.[35]

Mary Putnam very much enjoyed the intellectual atmosphere in the Reclus household, not only among the family members but with their frequent visitors as well. The Reclus, like so many French intellectuals of that era, maintained a salon and also took part in other salons. On Monday evenings their apartment was filled with friends and new acquaintances, usually people of the same political persuasion, all "démoc-soc" ("démocrate-socialiste") in the parlance of the day. To these meetings came émigré revolutionaries—Russians, Poles, Italians, and Spaniards—and all sorts of homegrown socialists, feminists, and others of liberal leanings. Paul Reclus, Elie's son, even told of a Basuto chief who entertained the gathering one evening by singing a song of his homeland. As a young man, Gabriel Monod, later to become a famous historian, admired the Reclus brothers but could not follow them all the way on the road to the left. As Monod wrote to a friend in 1868: "It is a rather interesting company that meets at the

Reclus' home, but I don't feel at all at ease among such dogmatic minds, who want to apply immediately institutions suitable only for perfect people, and who never dream of adapting the forms to facts." This from the future son-in-law of Alexander Herzen![36]

During the 1860s, when Elie and Elisée were still finding their way politically, they met a wide variety of people—among them Michelet, Blanqui, Clemenceau, and Proudhon—but the alliances were not enduring. They even briefly experimented with Free Masonry. The most important contact of that period was with Bakunin, from whom Elie managed to keep a safe distance but whose embrace was to turn Elisée in an increasingly radical direction. Actually, it is not unreasonable to believe that the Reclus brothers might have maintained their republican socialist ideas and not crossed over into true anarchism if it had not been for the great *débâcle*, or what Victor Hugo called "l'année terrible"—the terrible year of 1870-1871—when many French republicans, the Reclus brothers among them, were simply swept off by the radical tide and could not thereafter return to the safety of their previous port.

Chapter Four

The Debacle

The year beginning in July 1870 was indeed a terrible year for France and especially for Paris, but for Elisée Reclus it was the beginning of almost *deux années terribles.*

The summer of 1870 started out on a happy note for Elisée when he married his old friend Rosalie ("Fanny") L'Herminez, whom he had known since his sojourn in London in 1852. One of three daughters of an impoverished French family living in London, Fanny had not lost touch with Elisée, and she was working as a tutor in London when he asked her to marry him. They were married at Vascoeuil on 26 June 1870, without benefit of clergy or magistrate ("united by our free choice"). Elisée's wills, of which he made two in the two months following his marriage, were models of simplicity, the first saying merely, "I leave to Fanny L'Herminez all that the law allows me to leave her," and the second added a few words that might have had some legal significance: "I leave as a legacy to Rosalie L'Herminez, known as Fanny Reclus, all that the law allows me to leave her."[1]

Unfortunately, the happy couple were often separated during their three and a half years of marriage, suffering, like so many Parisians, the disastrous effects of the ill-considered war that Napoleon III contracted in mid-July. For the Emperor, the war was to end six weeks later at Sedan, but his erstwhile subjects, especially the inhabitants of the City of Light, had to face privation and destruction for several months because of the German war and the civil war that followed. As a socialist republican Elisée Reclus could hardly lament the fall of the Second Empire, even with the timid and tardy reforms

Napoleon III had initiated in 1869, but, as a patriotic Frenchman with a special fondness for his adopted city, Reclus could scarcely appreciate the foreign invasion that brought an end to the Empire. He certainly did not welcome the Germans as liberators, but neither could he muster the fierce invective that many of his countrymen were to use against the enemy.

With the outbreak of war Fanny was packed off to the provinces to be with the children, while Elisée remained in Paris with Elie and Noémi. Mary Putnam resided with them while continuing her medical studies, and it is to her that we owe some of the best accounts of the Reclus' life during the suspenseful days of the Siege.

When the first news of defeat came to Paris in August, Napoleon III seemed to fear the urban proletariat as much as he did the Germans. In a letter to her mother dated 14 August 1870 Mary Putnam said:

> I went upon the boulevards in the evening with Elie Reclus, and it was curious to see the soldiers stationed with arms, ready to fire upon the people. There was much more fear of an insurrection at Paris than of the enemy, and the government, which strips the hospitals even of their internes, does not hesitate to keep thirty or forty thousand soldiers at the capital, without speaking of the policemen, instead of sending them to the frontier, where there is the most urgent necessity to mass troops.

> It is really ridiculous to see how many people, who submit without a murmur to this outrage of the government upon two nationalities, and allow themselves to be robbed, ruined and broken-hearted by such an atrocious war, still keep up the old cry, "May Heaven preserve us from the Socialists. They are coming to destroy our property, our sacred property." It is enough to make one sick.[2]

By mid-September, after the establishment of the Republic, Mary Putnam was willing to play an active rôle in the war.

> My interest is immense in the events that are passing, especially since the Republic, and as far as I myself am concerned, I feel really quite ready to die in its defense, especially if in doing so I could help the Reclus. I probably shall not do so, however, in the first place because I feel that I owe myself as much as possible to you [mother]; in the next, because as yet there is no way clear by which I could serve the

Republic, either living or dying. I inquired yesterday at the Ambulance Society if there were any place, but they have already 4000 more names than places; so I went back and dug at my thesis, and probably shall stay there, unless Elisée Reclus is wounded on the ramparts.[3]

Elisée did not stand so much chance of being wounded on the ramparts as he did of contracting a cold or rheumatism from sleeping out-of-doors. He applied to Nadar, the famous photographer and balloonist, to be considered for entry into the balloon corps.

I beg you to let me know when and where I will be able to meet you to receive your instructions and begin my studies. I believe that I shall be able to be useful to you. To the advantage of being "heavier than air" I join that of being a geographer and a bit of a meteorologist. Moreover, I have the will.[4]

Although Nadar warned Reclus of the dangers, health hazards, and lack of pay, he remained undaunted and joined anyway. It is not known what Reclus's duties were, but apparently he did not take part in an actual balloon ascension. Although later writers, including Kropotkin, have credited Reclus with a major rôle in the balloon corps as well as in the establishment of the pigeon post, documentation of such grandiose claims is lacking. Reclus enrolled in a National Guard company in October, and he found it a great hardship to sleep on the ground, which probably accounts for his becoming immobilized with a severe cold in November. Although he wanted to serve the Republic, his heart simply was not in soldiering. Mary Putnam described him coming "downstairs, holding his rifle upside down" and concluded that "he was not, probably, a very serviceable solider."[5]

During the long Prussian siege, food supplies in Paris naturally grew sparse, but, although some of the desperate resorted to eating dogs, cats, rats, and even zoo elephants, the Reclus did not suffer great hardships because they had always known how to make do with little. Elisée had for some time been moving towards a completely vegetarian diet, and his wartime experiences strengthened his resolve. At the end of the Siege, Mary Putnam said, in a letter to her father,

I have never been ill, nor suffered from hunger, but, of course, our fare has not been delicate. Those who had made provisions for them-

selves, or who chose to spend a great deal of money for the gratifica-
tion of their stomach, and had it to spend, could always get along
pretty well. But my solidarity with the Reclus naturally prevented me
from making my "pile" apart, as I could not make it for them.[6]

Mary Putnam stayed with the Reclus until the German bombardment of Paris, which began on 5 January 1871 and continued for
three weeks. "We were obliged to decamp," she said, "for the bombs
rained around our house . . . I stayed with some friends in the north of
Paris,—the Reclus were dispersed in different directions." Mary Putnam apparently never saw Elisée Reclus again, although she was very
much concerned with his fate when he was imprisoned later in the
year. She and other members of her family maintained contact with
some of the other Reclus, especially Elie and his son Paul, for the next
thirty-five years, until her death.[7]

After the armistice of 27 January 1871 there was a great exodus
from Paris, especially of the bourgeois elements of the National Guard,
thus leaving the leftist republicans in a stronger position. They were
soon disabused of their hope of obtaining power in postwar France,
however, when they were overwhelmed by the conservative voters of
rural France in the national elections of 8 February 1871. The radical
element in Paris was convincingly crushed by the way in which the
election was conducted. Assembly seats were not awarded by arrondissement but were given to those who got the largest number of votes
in a city-wide election. Elisée and Elie were both candidates, but they
attracted few votes, and it is said that ballots marked simply "E.
Reclus" were voided. Elisée traveled to southwestern France in the
hope of presenting himself as a candidate for the Assembly from
Basse-Pyrénées, but he was disqualified because his papers were not
filed in time. These were his only attempts to enter politics, an area
that he subsequently regarded as *infra dignitatem*. After he was amnes-
tied in 1879, Reclus's name was proposed for a municipal council seat
in one of the arrondissements of Paris, but he graciously declined,
urging his well-wishers to support an honest workingman and not to
vote for one of the established politicians.[8]

After his futile attempt to spread republican propaganda in
southwestern France, Reclus returned to Paris in time to witness the
ultimate disgrace of the city before the victorious Prussians. The

republican ideals which had sustained the Parisians during the Siege seemed to be ignored. When the new government established itself at Versailles and announced some repressive measures aimed at the working class, such as the necessity of paying immediately rents and debts that had accumulated during the war, rebellion became inevitable. With his characteristic optimism, Reclus heralded the establishment of the Paris Commune on 18 March 1871. He naively believed that it would be the earnest of a federation of French communes, and he even made plans to take up residence in the Parisian suburb of Meudon. On the twenty-seventh of March he wrote to Alfred Dumesnil:

> It seems to me that the 18th of March is the greatest date in the history of France, after the 10th of August. It is at one and the same time the triumph of the Worker's Republic and the inauguration of the Communal Federation. Intellectual and moral progress had been immense since a change of this scope could be brought off almost peacefully.[9]

The period of euphoria was to end abruptly, for Thiers was amassing troops at Versailles, and the Civil War broke on 2 April, having been forced by a virtual declaration of war by Thiers. For Elisée Reclus the war was over almost before it began. Whether by choice or lack of opportunity, he did not have a position of leadership in the Commune but found himself again a regular soldier ("simple garde") in the National Guard, or as they were now called, the Fédérés, the Commune's army. He joined the 119th Batallion along with his younger brother Paul, who was still pursuing his studies at the Ecole de Médecine. Elie was not able to join the soldiery because of an old hand injury, but he was made Director of the Bibliothèque Nationale instead.

Immediately after the outbreak of hostilities the Commune decided to launch a major attack, later referred to simply as "The Offensive" because it was the only such action during the Civil War. The Fédérés marched on Versailles, with one wing, including the 119th Batallion, taking up quarters on the plateau of Châtillon. There they were quickly encircled by the Versailles troops on the early morning of 4 April and were told that all lives would be spared if they laid down their arms. Thus it happened that Elisée Reclus, who was probably still carrying a rifle incorrectly, gave it up without firing a shot. Brother Paul had remained in Paris and so was not present at Châtillon. The commander of the communards, Duval, was summarily executed along with former French army soldiers, recognizable by their uniforms, and the remainder, including Reclus, were marched off to prison. Along the way they were harangued and beaten by civilian spectators, whom Lissagaray derisively characterized as "the Parisian emigration—petty bureaucrats, elegant ladies, women of good breeding and street-walkers, jackals and hyenas." Thus began eleven months of imprisonment for Elisée Reclus, an experience which completed his radicalization and made it quite impossible for him ever to "work within the system" again.[10]

Although Elisée's contribution to the Commune was almost nil, his reminiscences of the Offensive and his subsequent imprisonment were useful to Lissagaray and other writers. When he was asked to write an article on the Commune in 1877, he said that it would be impossible for him to do so because his rôle as an actor and spectator was so minimal that it really did not count for anything, and he added, "To improvise what one does not know is a bad thing." However, twenty years later, when the editors of La Revue Blanche sent questionnaires about the Commune to numerous writers ("publicistes"), communards, and members of the Versailles army, Elisée Reclus, who was identified as a "publiciste" and not as a communard, gave a most interesting reply. He declared that since he had only been a simple guardsman and had been captured in the early part of the fighting, he only knew the Commune by hearsay and by later study of the events. In the first years after 1871 Reclus was unwilling to say anything bad about the Commune, but by the end of the century he thought it was time to tell the truth. He then stated his opinion that the military organization of the Commune was as grotesque and worthless as it had been during the Siege under Trochu: "the proclamations were as bombastic, the disorder as great, and the acts were as ridiculous." He thought that the officials of the Commune were honest but inept. They erred particularly in their retention of the old bureaucratic forms. They wrote diplomatic notes in a style that Metternichs and Talleyrands would have approved. They understood nothing of the revolutionary movement which had carried them to the Hôtel de Ville. But, he concluded,

> The Commune continues to inspire all those, in France and in the entire world, who wish to continue the struggle for a new society in which there will be neither masters by birth, title, or money, nor servants by origin, caste, or salary. Everywhere the word "Commune" has been understood in the largest sense, as referring to a new humanity, formed of free and equal companions, ignoring the existence of old boundaries and helping each other in peace from one end of the world to the other.[11]

In 1898 Reclus recalled his capture at Châtillon as one of the most glorious moments of his life. The bitterness of defeat was compensated by the mysterious and profound joy of having acted "according to my heart and will, of having been myself, despite men and destiny." He remembered the strong feelings of the Versailles troops against Paris and Parisians. Just as dawn was spreading over the city ("Never had the beautiful city, the city of revolutions, appeared more lovely to me"), a somber officer, "probably a country squire who had been raised by Jesuits," said, "You see your Paris! Well, there will not be a stone left standing!"[12] This prophecy was not borne out, for although the killing and destruction lasted almost two more months, the great city was to be repaired amazingly rapidly, at least physically, although the psychical wounds have not completely healed even after a century.

Another interesting commentary on the Commune was Elie Reclus's posthumous book *La Commune de Paris au jour le jour*. In his characteristic self-effacing manner, Elie denied playing a rôle of any significance whatsoever. He claimed to be only an observer, only "a thermometer hung in the corner," not one of the main actors or even a confidant of them. Elie was the Director of the Bibliothèque Nationale for twenty-four days (29 April to 23 May 1871), during which time he was as much concerned with possible depredations by fellow communards as by the enemy. His wife Noémi was a member of a commission to organize the teaching of girls during the Commune. Elie was not in his office on the day that the Versailles troops reached the Bibliothèque Nationale, and so he was able to elude capture, staying first with friends in Paris and then later making his way to Zurich by way of Italy. He was safely out of the country by the time he was sentenced to deportation to New Caledonia on 6 October 1871.[13]

After being captured at Châtillon Elisée began an imprisonment which lasted for more than eleven months and took him to fourteen prisons. The *fédérés* were taken first to Satory and then transferred to Brest. The passage to Brest was made under horrible conditions: a thirty-one-hour journey in an airless, filthy cattle car into which forty men were packed with almost no food. The prisoners were pleasantly surprised by the humane treatment accorded them by navy personnel in Brest. Reclus spent about six months in various prisons in Brest, particularly those of Quélern and Tréberon, before being taken to Versailles to stand trial. While in Brest he busied himself with writing and with various chores: teaching some illiterate prisoners how to read and write, conducting classes in English, Spanish, geography, science, and history, setting up a small library, and learning Dutch and Flemish. In a letter to Fanny dated 8 June 1871, he said that he was spending much of his time taking notes for a future work, *Le sol et les races* (presumably the *Nouvelle géographie universelle*), and also for "a purely literary little work," which must have been the *Histoire d'une montagne*, the companion-piece to the *Histoire d'un ruisseau*. Fanny visited Elisée once in prison in Brest. Apparently he was able to receive gifts of books, food, and money from his family. His parents sent him books, fruit, and chocolates in May; sister Loïs sent candy; and brother-in-law Pierre Faure sent a case of wine, most of which Elisée gave to the other prisoners because he preferred to drink tea. In August he told his Vascoeuil relatives that he had no need of money. He then had ninety-seven francs, and, since the prisoners were only allowed two francs pocket money a week, he had enough for forty-eight weeks.[14]

His publishers maintained contact with him during his incarceration. P.–J. Hetzel, who himself had been an exile from 1852 to 1860, offered every encouragement, moral and material, to Reclus. Hetzel continued to send Fanny an advance of two hundred francs a month for *Histoire d'une montagne*, even though Elisée was not able to complete the manuscript in Brest as he had hoped. Hetzel wanted him to translate some books, and Reclus offered to translate anything in German, English, Spanish, or Italian, but apparently he never undertook any such translations. While in prison he corrected the proofs for the second edition of *La Terre* and also finished writing the second volume of the abridgement, *Les Phénomènes terrestres*. To Emile

Templier of Hachette he proposed a geographical encyclopedia which would be published in little instalments each costing three or four sous. This proposal was a continuation of a plan which had been brewing in Reclus's mind since 1868 and which was to become the *Nouvelle géographie universelle*. The Paris Geographical Society was active in petitioning for his release, along with numerous friends, French and American. He was invited to attend the First International Geographical Congress, which met in Antwerp in August 1871, but his hosts in Brest would not let him go.[15]

Thus one can see that Reclus's imprisonment in Brest, while disagreeable in that it kept him from intimate contact with his friends and loved ones, was a fairly productive period for contemplation and writing. He was not so fortunate in the next five months that he spent in prisons in and around Paris during his trial and hearings. In late October 1871 he was moved from Brest to Paris to stand trial before the seventh Council of War at St. Germain. By a vote of five to two the Council sentenced him to "déportation simple," which meant deportation to New Caledonia, on 15 November 1871. Reclus's friends and relatives had not expected such a harsh sentence; indeed, they had expected a vote of acquittal; and the reasons which lay behind the vote will be obscure to me until I examine the original documents, which might be found in the archives in Vincennes. Nadar described Reclus at St. Germain as "a Christ before the praetorium . . . disdaining to defend himself, resenting, rejecting all palliation." William Huntington, the American newspaperman, in a letter to John Bigelow on 26 November 1871, said:

> Reclus' sentence is generally regarded as a very severe one. He behaved in a simple manly way before his judges, who themselves recognized his intellectual, literary, and moral worth, and made that apparently the chief reason for punishing him so heavily.[16]

Deportation to New Caledonia does, indeed, seem to have been a harsh sentence, especially to modern readers, but it was applied to large numbers of communards whose "crimes" were no greater than that of Elisée Reclus, and so I find it difficult to believe that he was singled out for persecution because of his moral superiority to those who were judging him. Surely the judges would not admit their inferiority. Rather, I think that it was Reclus's obdurate stance, his refusal to abjure future political activity, that made it difficult for his captors to turn him loose. His friends and professional colleagues kept up the pressure on various arms of the government to soften his sentence or to overturn it altogether. The Paris Geographical Society was especially faithful in seeking the release of its member. On 30 December 1871 a group of sixty English savants petitioned the French government to commute Reclus's sentence to simple banishment. A persistent story has it that Charles Darwin was among the signers, but his name does not appear on the petition. The English petition was not the work of any society or group but of an individual, Eugene Oswald, a German émigré living in London who had met the Reclus brothers through the Dumesnils. Oswald knew Elie, "a scholarly, firm, and reliable man," better than Elisée, but he leaped to the latter's defense when deportation to New Caledonia was imminent. As Oswald tells it,

> I consulted my experienced friend, Mr. Malcolm Ludlow, and we resolved to draw up in favour of the condemned man, and present to Monsieur Thiers, as the then head of the Government, a non-political petition in the interest of Science. Rapid action was necessary. Ludlow drew up our petition in a firm and dignified way, and I tramped the town for some days collecting signatures, chiefly among the heads of the learned societies of England. . . . Only occasionally had I to employ persuasion; any political bias was avoided, the man's scientific importance placed in the forefront. In about four days I gained about a hundred excellent signatures.[17]

Pressure was mounting for the commutation of Reclus's sentence. On 14 January 1872 Huntington reported to Bigelow that the sentence had been changed from deportation to banishment. "I suspect it is true or will be," he said, "though I find no official statement of the fact." On 1 February 1872 Reclus wrote to his parents from the Maison de Correction at Versailles saying, "It has been announced to me today in a manner 'not yet official' but as a certain fact that my punishment has been changed to banishment." The official announcement was not long in coming. On the third of February was issued the notice of commutation from "simple deportation" to "ten years of banishment, with loss of civil rights." There remained the choice of a country of exile. Although Reclus had previously considered several possibilities—the United States, Italy, Spain, Malta, and Switzerland—he now opted for the last-named because of Elie's presence there. In early March he was taken by rail to Pontarlier, about twenty miles from the Swiss border, and there kept an agonizing four days more in a local jail until 14 March 1872 when he was transported in a prison van to the border and turned loose. "Today I touched the free earth," he said in a note written that day en route to Basel and thence to Zurich. His incarceration had been surprisingly productive as far as his writing was concerned, and now he had a fierce determination not only to produce the monumental geography which he had long envisioned but also to work for the betterment of mankind, particularly through the dissemination of anarchistic ideals.[18]

Chapter Five

The New Geography

Elie had chosen to go to Zurich in 1871 because he thought it offered superior facilities for the education of his two sons, and it was his presence that attracted Elisée to Switzerland instead of to other possible countries of exile. After a joyful reunion, Elisée did not long remain in Zurich but moved instead to Lugano, near the Italian border, to which he was attracted, as he said to John Bigelow, by the climate, location on "the free soil of Switzerland," and proximity to Milan. He told Bigelow that he owed his freedom in great part to the action of his friends, especially in England, which attests to the efficacy of Oswald's campaign. He was most pleased to hear from Bigelow and to receive his gift of 125 German thalers, which he kept, not for himself but in the event he could use it to help some less fortunate comrades.[1]

Another reason for taking up residence at Lugano might have been the fact that Bakunin was then living in nearby Locarno, on Lake Maggiore. Indeed, Bakunin later moved to Lugano, but only after Reclus left. After the abortive revolution at Lyon in September 1870 and his subsequent propaganda defeat by Marx, Bakunin had little influence and few friends remaining. Switzerland became the headquarters of the anarchist movement, but the center was in the Jura and not in Italian Switzerland. The leftists of Geneva tended to side with Marx and despised the communards, who allied themselves with Bakunin and the Jurassiens. Bakunin had his greatest support in Switzerland, Spain, and Italy. Mediterranean lands seemed to favor Bakuninism as against Marxism, and in the Jura anarchism seemed to

[69]

take root quite naturally among the independent and intelligent watchmakers, who were, in Bakunin's words, "the last Mohicans of the late International."[2]

Bakunin had very high regard for the Reclus brothers, even though Elie fell away from him after 1869. In 1871 Bakunin described them in the following flattering portrait:

> . . .The two Reclus brothers, two scholars and at the same time the most modest, noblest, most disinterested, purest, and most dedicated men I have met in my life. If Mazzini had known them as I do, he would perhaps have convinced himself that one can be profoundly religious, all while professing atheism.
>
> United in principle, we are very often—nearly always—separated on the question of the realization of principles. They also . . . believed, at least they did two years ago, in the possibility of reconciling the interests of the bourgeoisie with the legitimate claims of the proletariat.[3]

In 1873 Bakunin described Elisée to Louis Pindy, a communard, as "a model man . . . noble, simple, and so modest, so forgetful of himself . . . He is a valuable friend, very reliable, very serious, very sincere, and completely our own," and added that he could forgive Reclus's comparative deficiency in revolutionary zeal. Very different in personality, Reclus and Bakunin nevertheless remained close. Max Nettlau, biographer of both men, has referred to their relationship as that of brothers and not of master and disciple. Bakunin despised authoritarianism, and since he regarded Germans and Jews as the most authoritarian peoples in the world, he had a special hatred for Karl Marx. Reclus could not be a hater, and, although his anarchism was very different from Marx's state socialism, he could not muster rancor against him. Marx's Germanic style found favor in northern Europe, Bakunin's Slavic style in southern Europe, and Reclus was attracted to the latter. Reclus remained loyal to Bakunin right up to the time of the latter's death on 1 July 1876, and he delivered a graveside eulogy two days later.[4]

Neither in his politics nor in his geography did Reclus engage in the great revilement that his countrymen unleashed against the Germans after the Franco–Prussian War. Although the quality of French science, including geography, vis-à-vis German science had been declining since the 1840s, the gap became especially apparent with the rapid success of the Germans in the war, and the French sought to redress the balance by imitating their foe. There was a great expansion of the French scientific establishment after 1871 and a concomitant reform of education at the university and secondary levels, but the French educational system did not undergo the complete scientization that occurred in Germany. The French described German science as being characterized by *l'esprit géométrique* and their own by *l'esprit de finesse,* the implication of course being that French scientists were more concerned with quality than with mere measurement. Interestingly, these same phrases have been used in the twentieth century in making the distinction between the French style of geography and the German (and American) style. In any event, Elisée Reclus always insisted that science is international, never simply a national enterprise, so that to speak of "French science" or "German science" or "Italian science" is only invidious. He derided those who would advocate protectionism for productions of the mind as they would for turnips or cotton goods. In 1891 Reclus proposed an alliance with the German geographer Albrecht Penck in order to "overcome that abominable impasse of 'French science and German science,' " but Penck did not respond.[5]

Virtually the only instances of anti-German feeling that I can find in Reclus's writings are in a few of the anonymous columns on "Géographie générale" that he wrote for Léon Gambetta's newspaper, *La République Française,* between February 1872 (when he was still in prison) and January 1875, but one cannot be absolutely certain of Reclus's authorship of every one of the articles. It is said that Gambetta, "commis voyageur" (traveling salesman) of the Republic, used the newspaper as "the instrument of a shadow cabinet" and that "his principal followers were each given one specialty to write about . . . proposing policies they would be ready to implement," but Reclus was not close to Gambetta, and he certainly could not have been considered as a potential cabinet member. As an exiled communard, he would have been too much of a maverick even for Gambetta's bizarre tastes.[6]

Geography generally profited from the educational reforms of the Third Republic. It was one of the chief beneficiaries of the reduction of

[71]

classical languages in the curriculum. In most European countries and in the United States the modernization or professionalization of geography began at nearly the same time—in the 1870s and 1880s—and in France the New Geography was initiated by scholars who were trained as historians, particularly Paul Vidal de la Blache, whose long tenure at the Ecole Normale Supérieure (1877-1898) and the Sorbonne (1898-1909) caused him to become known as the Father of Modern Geography in France. That title might have gone to Elisée Reclus, but his lack of a doctorate prevented him from being a professor in a French university, and the New Geography was university centered. However, the public success of Reclus's writings undoubtedly eased the way for Vidal and his students, and they were grateful that they could operate in the favorable milieu that he did so much to create.[7]

Reclus's international reputation had begun in 1868 with the publication of the first volume of *La Terre,* but his everlasting fame results from the publication of the nineteen-volume encyclopedic work, *Nouvelle géographie universelle,* which appeared in weekly instalments over a period of nineteen years. The plan of this monumental work was conceived even before *La Terre* was finished, and he might have been able to work on it even during his imprisonment in Brest, but it began in earnest when he reached Lugano. During his visit to Zurich in March 1872, Reclus drew up a "Plan de géographie descriptive," which he sent to Emile Templier of the Hachette publishing house. The *Géographie descriptive* would be a continuation of two works, *La Terre* and *L'Homme,* the latter not yet written. With the general aspects of cosmography, geology, physical geography, anthropology, and linguistics covered by *La Terre* and *L'Homme,* Reclus could then devote the *Géographie descriptive* to description of the various countries of the world. The initial plan called for five or six volumes about the size of those of *La Terre.* As for the order in which he would treat the regions of the world, Reclus said that if he were to follow an historical order, Bactria should perhaps be first, a geographical order might call for the great plateaus of Asia to be given the position of primacy, and a geological order might have volume one commence with Australia. Instead, he wished to follow the habitual order and begin with Europe, and pride of place would go to Switzerland, as the orographic center of the continent. The description of each country would begin with a brief statement of its natural or political limits, followed by a discussion of the history of exploration, a description of the physical geography according to the order established in *La Terre* (relief features, hydrography, weather and climate, flora and fauna), and finally the human geography. After treating the various ethnic groups and their special characteristics, he would proceed to what he called the "statistical" part: the discussion of population, economic activities, political constitution, and the "intellectual and moral state" of the people. Administrative subdivisions, a major subject in conventional geographies, would be given short shrift and perhaps even relegated to an appendix. A large number of maps would be needed. Reclus hoped that the work would be liberally illustrated with thirty to forty colored maps and perhaps 700 additional diagrams, pictures, and black-and-white maps. [8]

Emile Templier wrote to Reclus on 27 May 1872 to say that Hachette had that day decided to undertake the publication of the proposed work. Templier accepted the plan but thought that Reclus was giving too much emphasis to "ancient geography" (history of discovery and exploration, origin and progress of the races), which is "more in the realm of historical geography than descriptive geography." He also felt that the subject of administrative subdivisions should not be relegated to appendixes. Templier hoped that the work could be published in weekly instalments (livraisons) of sixteen pages each over a period of four years. He agreed to pay Reclus six hundred francs a month during that time (a minimum of 28,800 francs altogether) and to establish a royalty of two centimes for each livraison sold. The publication would commence sometime in 1873. [9]

Hachette subsequently drew up a contract to formalize the agreement. The contract, signed by Reclus on 8 July 1872, called for a "Géographie descriptive et statistique" which would cover all the countries of the world and would contain around seven million letters (*sic*). The work would consist of five or six volumes, to appear in 200 to 210 weekly instalments of about 35,000 letters each. The contract called for the six hundred francs as a monthly stipend that Templier had suggested. [10]

Reclus soon sent Templier his description of Switzerland, which was to be the first part of the initial volume. Templier read it over and returned it on 31 August 1872 with the following comments:

Your work appeared to me to be clear, precise, well organized, and
. . . completely worthy of you; but it is very necessary that I confess to
you that I have not found there what I was looking for above all. The
general concepts, the overall views, the enthusiasm for the great
spectacles of nature—in a word, all that can give charm, interest, and
life to a book of geography is completely lacking there.

No one is better able than you to animate such a subject; in not doing
so, you have certainly acted with premeditation. I have an idea of the
reasons which could have led you to adopt such a rigorously didactic
form, but I do not think that the reasons are sufficiently strong to
prevail against those that I am going to submit to you:

We are undertaking a very large and expensive publication, and we
need, in order to carry it through successfully, to attract the largest
possible clientele. The support of society folk is no less necessary to us
than that of specialists. Now, if you content yourself with giving us a
book of information, the scholars alone will consult you, and we shall
lose nine prospective readers out of ten.

It is not merely a school book or reference work that I wanted from
you. It is a *literary* book, a sort of poem of which the Earth is the hero. I
would wish that you might give a large place to the picturesque
description of each country, and as far as it would be possible without
upsetting your plans, a large part also to man, his customs, and his
works. I should like to see you enclose in that description the grand
results of statistics and to slide over the details.[11]

Reclus was miffed by Templier's statement that general concepts
were completely lacking in his work. He reiterated his intention of
writing a very sober and almost purely didactic work but one which
would be "less arid than the majority of general geographies known to
me." He then suggested some minor modifications in his plan "in order
to give my work more interest and literary charm, without diminishing
its scientific value." He asked Templier to let him know right away if
these suggestions were agreeable to him in order that he might recast
the first livraison and use it as a model for the eight to ten other
livaisons which he had already prepared. "Time is precious and life is
short." Templier approved and returned the manuscript on Switzerland to Elisée through Onésime, who was his brother's Paris inter-
mediary with Hachette.[12]

Just before Christmas 1872 Reclus sent Templier a manuscript of 171 pages, of which 100 were the pages on Switzerland that he had rewritten. Templier read the manuscript in the course of the next month and again offered several suggestions to Reclus about his writing style:

> I must admit that I am lost in that long description of mountains, which has appeared even to you to be a little monotonous; the main topics have disappeared in the midst of details. . . . In my opinion it is necessary to make the reader's task easier and more profitable; and you could do it without difficulty . . .
>
> I also wish to speak to you of a lacuna which struck me and which would be regrettable, in my opinion.—You do not discuss *towns*. These dwelling-places of contemporary people nevertheless have as great geographical importance as the Neolithic lake dwellings.[13]

Reclus and Templier soon came to an agreement about such matters, and by June 1873 Templier no longer felt that it would be necessary to read critically every page of Reclus's material. "Now that we are in accord on the method and spirit of your work, you no longer have to be worried about my opinion on the manuscripts that you send me; it is not even certain that I will read them all before publication."[14]

Many factors were to conspire to delay the publication of the first livraisons. Reclus insisted on having a large number of maps, and it was very hard to find and keep a good cartographer. The cartographers in Paris were overloaded with work, and Reclus despaired of finding a decent draftsman nearer to Lugano than Bern or even Vienna. Reclus was also working on other writing projects, and he took several weeks off for a visit to the Vienna Exposition in 1873 which he followed with a long excursion into Transylvania. The greatest interruption to his work schedule was the sudden death of his wife Fanny in 1874 and his subsequent departure from Lugano.

Fanny had been a perfect wife to Elisée, and so when she died of puerperal fever on 14 February 1874 along with their newborn son Jacques, he was grief stricken. She was a different type from his first wife, Clarisse—"the latter, all gentleness; the former all will," according to nephew Paul. "Her power for work was unimaginable," said

[75]

Paul, "and her line of conduct inflexible. In many ways Elisée and Fanny found themselves at one; she was a woman to his mind." Beginning in 1872 and even for twenty years after Fanny's death, Elisée often signed his name "Elisée Fanny Reclus," "Elisée F. Reclus," or "Elisée FReclus." Although they were married for only three and a half years, during which Elisée spent eleven months in prison, and although their marriage had been initially based more on practical considerations than on romantic feelings, they had come to love each other very deeply, and Elisée was badly shaken by Fanny's sudden demise. Fanny's mother, Mme. L'Herminez, had been living in Lugano with them, and she died of apoplexy on 7 July 1874, just before Elisée moved to a small village at the eastern end of Lake Geneva.[15]

Immediately after Fanny's death Elisée began to think seriously of leaving Lugano. His primary reasons for the move were his concern for the education of his daughters, whom Fanny had been tutoring, and also for the drafting and engraving of maps for his geographical encyclopedia. Elie and Noémi wanted him to return to Zurich, but he declined for several reasons. He did not want to move his children from the semi-Italian atmosphere of Lugano to the Germanic milieu of Zurich. Furthermore, Fanny had disliked Zurich, and Elisée retained a superstitious respect for her feelings. In a letter to Noémi in March he seemed to be leaning towards Italy, not to Milan which is situated "in a plain exposed to all the winds and which furthermore lacks the engraving establishment which I require," not to Florence which "is too far in every respect, too far from you and too far from Paris," but perhaps to "dull" Turin, or, better yet, to the more sheltered towns of Chieri or Moncalieri just to the south of Turin. In a letter to his brother-in-law Edouard Bouny dated 24 April 1874 Elisée said that he was then considering either Turin, Brussels, London, or a Swiss locality on the other side of the mountains. He summarily dismissed Brussels, "la ville de la plaine et des contrefaçons" (fakes), an interesting judgment in view of the fact that he was to spend the last eleven years of his life there. He was afraid that Turin's climate would be too extreme. He rather favored London, but his relatives tried to point out its disadvantages. As for Swiss towns, Montreux was a distinct possibility. A major concern was the education of Magali and Jeannie, and Elisée wanted to entrust them to the care either of his sister Noémi (Mme. Eugene Mangé) or to Fanny's sister Lily, who was then in

London. Lily was to come to Switzerland in late summer to take care of the girls, but in the meantime Elisée thought of asking his old friend Mme. Ermance Trigant-Beaumont to instruct them in natural history. Ermance Gonini, three years older than Elisée, had married a cousin of his mother in 1859. She was widowed within a year and left with a small fortune. After that her life intersected with Elisée's on several occasions—they were together in London in 1862 and in Hyères in 1863—so that it is not remarkable that Elisée, or someone in the family, thought of her as a possible tutor and companion for his children in 1874. Thus, in much the same way that Fanny came back into his life after the death of Clarisse, Ermance was to reappear and to fill a gap in the family circle. The practical advantage of association with Ermance became evident to Elisée, and they were married ("free of all religious or legal formality") in Elie's house in Zurich on 13 October 1875. Elisée moved first to La Tour de Peilz, a small village on the eastern end of Lake Geneva, in July 1874, just after his mother-in-law died, and then moved to nearby Vevey a year later.[16]

After Elisée married Ermance, she apparently paid the costs of housing and household expenses, thereby increasing his financial security at a time when he was getting a comfortable income from Hachette. Just before the marriage, Elisée started keeping an account book, which he maintained for almost thirty years, until his death. Although it is impossible to determine his net income after paying expenses incurred in connection with his writing, it would appear that he gained a rather good income, particularly from the *Nouvelle géographie universelle*, which yielded a steady monthly emolument right up to the beginning of 1905. I have calculated that, according to the figures in the account book, which might not be perfectly complete or accurate, Elisée Reclus received more than half a million francs from Hachette for the *NGU* to January 1905. This would mean an income of more than 16,000 francs a year—a handsome salary, especially if his wife was paying all the household expenses. However, Elisée never affected the bourgeois style; he always lived frugally, with little regard for creature comforts; and any temporary surplus was promptly disposed of in helping needy relatives and friends and in promoting the anarchist cause.[17]

Reclus's strict ethical code was manifested even in his vegetarianism. He had developed a repugnance to the slaughtering of animals when he was a small child, and long years of penury gave him abstemious habits. Elie was also a vegetarian, but he would eat meat or whatever was put before him so as not to offend his hosts. Characteristically, Elisée adopted vegetarianism more avidly than his brother, and during his last years he became a nibbler and did not usually partake of full meals. Fruit, nuts, bread, and little cakes were his favorite foods; they were perfectly suited to his spartan code and his sedentary life. Reclus's puritanical habits provoked admiration and a little amusement among his friends; Gustave Courbet, for example, said that Reclus lived on lentils and water and warned against accepting invitations to dine with him.

Courbet also said that Reclus was like a tailor, in that he rose at dawn, worked steadily through the day, and only put down his tools after the sun had set. Reclus's ability to write incessantly, not only in his study but wherever he happened to be, was noted by numerous contemporaries. Luigi Galleani, a friend of the Swiss period, said that Reclus would write for fourteen hours a day and then go for a walk at sundown or swim in Lake Geneva. Despite annoying touches of angina after 1880, Reclus was a hardy traveler and a dedicated believer in the benefits of physical fitness. At the age of fifty he engaged in some gymnastics with his future sons-in-law, and he liked to hike and play with children, even to the extent of climbing trees with them. He was fortunate in that the circumstances of his occupation and his family life allowed him to set his own schedule and to work single-mindedly toward his goals. How different his life would have been if he had been, say, a university professor with all the necessary distractions such a job entails.[18]

Like most monumental publishing ventures, the *Nouvelle géographie universelle* did not proceed according to schedule. Instead of five or six volumes to be published within four years from the time of the signing of the contract in July 1872, it actually ran to nineteen volumes, the last of which appeared in 1894. The first livraison, originally scheduled to appear in 1873, was published 7 May 1875, at which time it was thought that the complete set would run to 500 livraisons, or ten to twelve volumes. The final livraison of volume one was published a year later, and Hachette then announced the sale of the first volume (all the livraisons bound together). The first volume, which had originally been planned to begin with Switzerland, actually started with Greece and then proceeded westward through the Mediterranean to the Iberian Peninsula. Switzerland was finally included in volume three (Central Europe). Templier had ventured the opinion that France should be treated after Italy and before Spain, but instead it was given the whole of the second volume.[19]

Even on the eve of the publication of the first livraison, Templier and Reclus could not agree on a title for the work. Elisée had suggested *La Terre et l'homme*, but Templier thought it would be confusing to call Reclus's second work *La Terre* like the first, and this confusion might be reflected in weak sales. Onésime proposed *Les Deux-Mondes*, and Templier countered with *Les cinq parties du monde*. Still dissatisfied, he thought of *Le Monde où nous vivons* and then *Le Monde terrestre*, *Le Monde et ses peuples*, and *Le Monde et les hommes*. Elisée suggested the simple designation *Géographie descriptive et statistique*, but Templier did not like the word "statistique" and said that a substitute word denoting generality or universality should be sought. Thus the final compromise title became *Nouvelle géographie universelle*, indicating that this work would supersede Malte-Brun's outdated *Précis de géographie universelle*, and then the phrase *La Terre et les hommes* was added as the subtitle.[20]

After the first livraison appeared in May 1875, it was planned that the rest would be published at weekly intervals, every Saturday. This mode of publication caught the public fancy, and Hachette soon began printing 20,000 copies of each livraison. However, sales declined noticeably by the spring of 1876, and bookstores began to return large numbers of unsold copies. The publication of volume one—all the livraisons on Southern Europe bound together—was announced in the 17 June 1876 issue of the *Bibliographie de la France* at a price of thirty francs, which effectively kept the book out of the hands of workingmen. It is ironic that a man of the people would write books which would be owned and read mostly by the bourgeoisie and aristocrats. Reclus must have rationalized that the income, from whatever source, was helpful in advancing the anarchist cause and that the books, containing their subtle message of the brotherhood of man, were important in infiltrating and liberalizing the libraries of the well-to-do. In any event, the *Nouvelle géographie universelle* brought Reclus a comfortable income for over three decades and enabled him to order his life with a great deal of freedom.[21]

[79]

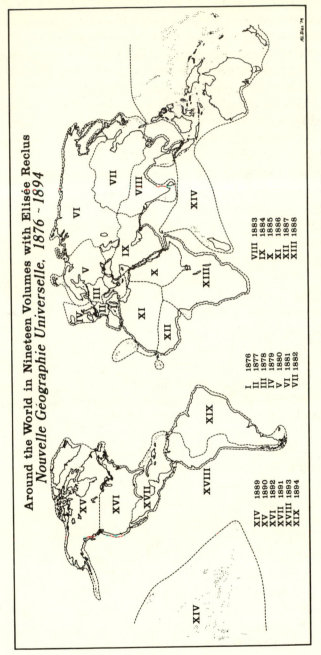

Around the World in Nineteen Volumes with Elisée Reclus
Nouvelle Géographie Universelle, 1876 ~ 1894

I	1876	VIII	1883
II	1877	IX	1884
III	1878	X	1885
IV	1879	XI	1886
V	1880	XII	1887
VI	1881	XIII	1888
VII	1882		

XIV	1889
XV	1890
XVI	1891
XVII	1892
XVIII	1893
XIX	1894

Map showing the areas covered by the nineteen volumes of Reclus's *Nouvelle géographie universelle*. Equal-area projection allows comparison of the sizes of the land areas treated in each volume.

[80]

In July 1876 Templier told Reclus that the first two volumes were running about one-third longer than planned. He asked Reclus to cut each volume down to about forty or forty-five livraisons. Hachette decided to publish two livraisons a week instead of one and hoped to start new volumes on January first of each year, presumably in order to have the completed volumes ready for sale during the holiday season as *étrennes*. In fact, it seems that few of the nineteen volumes of the *NGU* actually appeared at the end of the calendar year; most of them seemed to be finished around mid-year, but reprints were often released at year's end for the Christmas trade.[22]

The first volume covered Southern Europe with the exception of France, which was made the subject of volume two. After an introduction which was strongly reminiscent of the epilogue of *La Terre* ("Study with me this 'Beneficent Earth' which supports us all and on which it would be so good to live as brothers"), Reclus began with Greece and moved westward to the Iberian Peninsula. The organization of the volume is interesting, for Reclus developed a formulaic style here which he employed in all but two of the nineteen volumes, the exceptions being the volumes on France and the United States. His organizational unit was the nation-state, and the number of pages devoted to each nation was proportionate not only to its population but also to the amount of literature available in western European languages. (The number of pages also seemed to be related to the country's distance from France.) It was necessity, not desire, that caused Reclus to adopt a Eurocentric view, and he could not fairly be charged with ethnocentrism or racism, even by the extremely sensitive standards of the 1970s. Typically, Reclus's treatment of a country—a single chapter, if the country was small enough—would consist of an introductory section devoted to the history of exploration and early settlement, early ethnic distributions, place-name etymologies (often inaccurate), a section on physical description following the traditional (or what Reclus helped to become traditional) pattern of geology and landforms, hydrography, weather and climate, and flora and fauna, and then a longer section on human geography and regional description, finally ending with a dry accounting of present-day commercial and political data, which he might have appended somewhat grudgingly because of Templier's earlier criticism. Each volume was well illustrated with maps, sections, and pictures, often drawn from photographs. A welcome innovation was the inclusion of numerous city plans and other maps of rather large scale, often redrawn from existing topographic maps. Few of the illustrations were unique to this enterprise; most had been published elsewhere and were redrawn by Reclus's cartographers and the Hachette stable of illustrators. Reclus himself usually selected the maps and other illustrations to be copied, and this chore consumed a very large part of his time. He was particularly fortunate to have such an able cartographer as Charles Perron (1837-1909), who had been one of Bakunin's followers in the 1860s, as the principal draftsman for the *NGU* from volume three onward.[23]

The first volume found a very receptive audience, and the reviews were enthusiastically favorable, although not always uncritical. An anonymous reviewer in *L'Année géographique,* presumably Vivien de Saint-Martin, remarked particularly on Reclus's pleasing style and originality of method:

> The plan belongs uniquely to the author; it has nothing in common with that of previous works—with the *Précis* of Malte-Brun, for example, the most beautiful, most complete, and most learned general geography that we have had up to the present. The *Nouvelle géographie* of Mr. Reclus is, one can say, a suite of discourses on each of the countries and states described, discourses full of facts and a profound learning in that which touches the physical side of the subject, in particular, and the knowledge of populations, rather than a didactic description bolstered by proofs and supporting references. The author never cites his sources, no more than Buffon in the magnificent discourses which open each of the great divisions of his *Histoire Naturelle.* One can have some reservations about the absolute employment of this essentially literary method; but no one can deny that it is supported by a grand style and by splendid accessories, maps, plans, views, and typography. The prodigious success of this new work of the author of *La Terre* moreover proves that he has wonderfully grasped the taste and temperament of the general public.[24]

More penetrating were the reviews of Henri Gaidoz, an archeologist and mythologist, in the *Revue critique d'histoire et de la littérature* and *La Revue politique et littéraire.* In the former he was especially critical of the maps and of Reclus's style, which he considered more appropriate for a literary than a scientific work.

It would be better if the maps were less numerous and more carefully done. . . . It is impossible for us to accept as anything other than geographical caricatures the maps of "Relief de l'Europe" and of "Zone isothermique de l'Europe" . . . This criticism is addressed perhaps more to the publisher than to the author. But we must reproach the latter for trying to give his work an exclusively literary cachet. The style of M. Elisée Reclus is naturally vivid and especially attractive because one senses in it an original and personal thought: but M.R., falling under the influence of a celebrated journal where he has written much [*Revue des Deux Mondes*], readily allows himself to fall into writing phrases where vague amplifications take the place of facts, numbers, and dates. . . . M.R. has undoubtedly feared looking like a pedant; he has wanted to make a work of popularization and not a work of science; but . . . in avoiding the appearance of pedantry one risks giving oneself the appearance of frivolity.[25]

Gaidoz then critized the ethnographic map of Europe which appeared in the second livraison, showing his jealous concern for the representation of German-speaking communities.

M. R. has forgotten to show the Wends of Lusatia who form in the middle of the German population a group at least as important as the German colonies carefully indicated on the Volga. . . . M.R. counts 36,000,000 Frenchmen; we believe that figure lower than reality by nearly one million. On the contrary, he counts 52,000,000 "Germans and Swiss" (sic!) which is too high by several million. . . . The name Swiss as used by M.R. . . . has no more ethnographic significance than the name Austrian. The name Turanian is chimerical. One cannot speak of 'populations of Greco-Latin languages,' for the languages which are Latin are not Greek, and vice versa.[26]

Gaidoz praised Reclus's work generally and recommended it highly to the French public, who, he thought, had not been exposed to geographical works of high quality. In subsequent reviews of the *NGU* in *La Revue politique et littéraire,* he continued to praise Reclus for being a happy combination of scholar and writer, geographer and historian, and even writer and artist. When Gaidoz regretted that historical and political geography, "those two great sides of geography," were summarily treated, he was thinking of the arid lists of historical events and names of administrative divisions which were often included in geographical compendia. Gaidoz's sensitive antennae detected a tolerance towards the despotic rulers of the Ottoman Empire which Reclus would not have shown towards European monarchs. Actually, Reclus, although always fascinated by Turkey (his visit to Istanbul in 1883 marked his farthest eastward venture), was fully aware of its shortcomings. In an article in *La République Francaise* in 1872, he had said, "Of all the countries of Europe, Turkey is that which occupies the most fortunate situation and is most richly endowed by nature." Although he called Turkey "an empire where caprice is sovereign" in the first volume of the *NGU*, he retained an optimistic hope that Turkey would undergo regeneration which would put her on "a normal path of peaceful and continual progress." With ingenious logic, he even tried to show that the Christians within the Ottoman Empire could perhaps cope with Turkish rulers somewhat better than they might be able to do with European rulers.

> Fortunately, Turkish despotism is not an educated despotism, based on the knowledge of men and aiming methodically at their debasement. . . . Their domination is often violent and cruel but it is all exterior and does not reach the depths of the being. . . . [Non-Turkish minorities] can easily resist a domination which is capricious and devoid of plan.[27]

In any event, there is plenty of evidence to show that Reclus did not merit Gaidoz's charge of "turcophilisme." In a letter to his wife from the banks of the Dardanelles in 1883, Reclus said that "the Turks are the most backward people in the world because they are the most religious." It is difficult to imagine what index of religiosity he had in mind, but I think that this statement simply shows that Reclus had many of the normal foibles and prejudices; the public Reclus was more tolerant and optimistic than the private Reclus.[28]

Gaidoz had high praise for Reclus's volume on France—"one of the most beautiful of the serious gift-books of this year"—but he felt sad that "a man who has described France with so much knowledge and love should be . . . condemned to live outside of France." Reclus did indeed have a great love for his natal land, but this book was another example of the difference between his private thoughts and published statements. The unity and individuality of France, and consequently the integrity of her political frontiers, are age-old themes with French geographers, and Reclus's volume on France is cited as one of the monuments of that genre. In fact, in letters to Templier, Reclus said that the unity of France was exaggerated, and, as a good Gascon, he perceived great differences between the Nord and the Midi. Templier had disagreed, giving his opinion that France displays greater unity than any other country of Europe. In the *NGU*, Reclus appeased Templier by saying that France "is one of the countries which has the greatest national unity," but in his posthumously published *L'Homme et la terre* he said that "France as a whole is less unified than Germany and even Italy." In the *NGU* he repeated the statement that he had made in Joanne's *Dictionnaire* in 1864 to the effect that France "is distinguished among all the countries of Europe by the elegance and equilibrium of its forms." Although France's historical rôle had been disproportionately great, he modestly demurred from claiming moral superiority for the French as well.[29]

It is somewhat ironic, but not surprising, that Elisée Reclus took a more active interest in anarchist activities in the middle-1870s when his books promised him a steady and comfortable income and he settled down on the eastern shores of Lake Geneva with a wife of means. A powerful stimulus to this activity was his association with Peter Kropotkin, whom he first met in February 1877. Kropotkin, a Russian aristocrat, had traveled extensively and had ample field experience in physical geography and geology, but when he observed the poverty of the Finnish peasantry in 1871 he gave up his cherished ambition of obtaining the secretaryship of the Imperial Geographical Society in St. Petersburg and decided to devote his life to social action. Imprisoned for his new faith, he escaped to western Europe, first to Edinburgh where he met the brilliant Scottish polymath Patrick Geddes, and then to London where he met the geographer John Scott Keltie, editor of *Nature*, for whom he had already written some pieces. Keltie was the anonymous reviewer who praised Reclus's *NGU* lavishly in the London *Times*.

As George Woodcock and Ivan Avakumović have said in their biography of Kropotkin, his observations on glacial deposits in Finland marked "the virtual end of his career as an original geographer"—*i. e.*, as an explorer and field worker. However, he remained a writer and lecturer on geographical subjects for at least four decades, until his return to Russia in 1913. His background in geography was of vital significance to his development as an anarchist, just as in Reclus's case. Woodcock and Avakumović describe the connection in this way:

> His expeditions had developed in him resourcefulness, independence, and the understanding of men in widely different communities and circumstances, while the theoretical problems he encountered had matured his powers of thought and given him a feeling for scientific method which was to characterise his subsequent thinking, whether revolutionary or biological, sociological or ethical. Moreover, it is difficult to imagine a science better fitted than geography to lay the foundations of his very wide understanding of the sciences. . . . [30]

Kropotkin first came to Switzerland in December 1876 while working on the Russian and Siberian sections of a gazetteer edited by Keltie. He met Reclus in Vevey in February 1877, and the two geographers took an instant liking for each other. As Kropotkin reported that first meeting:

> I liked him a lot. We discussed much, and I was pleasantly surprised to find a true Socialist (I was a bit mistrustful about that because of his scholarliness). [31]

Although different in background, the two men were remarkably similar in character and in their zeal for human betterment. Their common base in geography gave them great breadth of vision and tolerance, according to Elisée's nephew Paul. Max Nettlau, who knew both Reclus and Kropotkin well, also pointed out the common elements in their intellectual development:

> Both tore themselves from milieux which had clung to them through birth and education, Reclus from the religious, Kropotkin from the aristocratic-military. . . . Both were led by the drive for knowledge and love of humanity towards the most comprehensive study of nature and man, not in order to specialize in a narrow area but to recognize on the basis of exact observations the way of social evolution and going from theory to practice to remove the obstacles to this evolution. . . . Both were early introduced to communism, Reclus to the idealizing communism of the early Christians and later persecuted religious sects, Kropotkin to the primitive communism of Russian peasants;

both got to know primitive people in their natural habitat (in South America and in Siberia, respectively) . . . Neither belonged to any School or Party.[32]

Nettlau also pointed out some differences between Reclus and Kropotkin, and he preferred the former's style. "Reclus always ranged a step higher, standing on a wider, higher platform" than Kropotkin. "To me," said Nettlau, "Kropotkin's Anarchism seems harder, less tolerant, more disposed to be practical; that of Reclus seems to be wider, wonderfully tolerant, uncompromising as well, based on a more humanitarian basis."[33]

Before his explusion from Switzerland in 1881, Kropotkin worked closely with Reclus and the Jurassiens. He warmly approved the lifeways of the Jura Swiss watchmakers, who lived and worked in model communities. Kropotkin and François Dumartheray founded an anarchist newspaper, Le Révolté, in Geneva in 1879, and the journal received every form of assistance, moral and material, from Elisée Reclus. He wrote for the newspaper and also gave it a monthly subsidy of a hundred francs for the fifteen years of its existence. After six years of publication in Geneva, Le Révolté moved to Paris in 1885 and was renamed La Révolte in 1887. It disappeared temporarily during the period of repression in 1894 and then reappeared the following year under the name Les Temps Nouveaux. Throughout its checkered history, despite changes of venue and editorship, the journal retained the spirit of moral indignation which was expressed so forcefully in the first issue: "We are rebels [révoltés]" against the present society, which "spells extreme misery for some . . . insolent opulence for others," with the aristocrats on top, the workers on bottom, and women in the worst position of all.[34]

Of special interest to geographers are two anonymous articles that Reclus wrote for Révolté/Révolte in 1887: "Les Produits de l'industrie" and "La Richesse et la misère." Here he asserted that human misery is due to the monstrous organization of present-day society and not to lack of resources, since the earth produces twice as much food and three times as many industrial products as are required. Reclus was opposed to the Neo-Malthusianism of many French anarchists, such as Paul Robin, because he believed that it was maldistribution of re- sources, rather than overpopulation, that was responsible for the world's ills. "Thanks to geography and statistics," said Reclus to his English friend Richard Heath in 1884, "[We know] that the resources of the Earth are amply sufficient so that everyone has enough to eat . . . It is in the name of science that we can say to the learned Malthus that he is wrong." First Reclus put strange words into Malthus's mouth—"At the great banquet table of present-day society there is no place for the poor"—and then he attacked them fiercely. Reducing the numbers would not assure adequate subsistence to the living. France's declining birthrate, a subject of much concern in late 19th-century France, was described by Reclus as a blessing: "If millions of young women and men had been added since 1870 to the decimated ranks of the present generation, what would the priests and politicans have done to them? May mothers refuse to conceive so long as the political and social milieu condemns them to give birth only to victims or executioners."[35]

Although the *Nouvelle géographie universelle* was a one-man effort in that it was entirely planned and written by Reclus, he had a considerable amount of help from Kropotkin and from a succession of secretaries—Gustave Lefrançais, Léon Metchnikoff, and Henri Sensine. Kropotkin was credited with rendering aid with the parts of volumes five and six that dealt with Finland, European Russia, and Siberia. While Kropotkin was in prison in Clairvaux, France from 1883 to 1886 his suffering was mitigated somewhat by the books and money which Reclus sent him. At Clairvaux Kropotkin also worked on a series of essays for which Reclus supplied the title *Paroles d'un révolté*. In fact, Reclus thought up the titles of Kropotkin's other French works as well—*La Conquête de pain*, *Autour d'une vie*, and *L'Entr'aide*. Kropotkin was fluent in French but preferred to write in Russian or English, while Reclus, who was fluent in English and German and knew several other languages, including Russian, fairly well, wrote (for publication) only in French.[36]

Reclus's first assistant in Switzerland was Gustave Lefrançais, a communard who returned to France after the amnesty of 1879. He was followed by Léon Metchnikoff, an erudite Russian-born orientalist who was able to give Reclus much help with the East Asian volume (volume seven) of the *NGU*. In 1883, with Reclus's backing, Metchnikoff became professor of geography in the University of Neuchâtel, but he continued to assist Reclus until his death in 1888. Metchnikoff wrote a great historical and geographical compendium on Japan, *L'Empire japonais* (Geneva, 1881), and also a posthumous work which is still of some interest almost ninety years later—*La Civilisation et les grands fleuves historiques*, a marvelous tour de force to which Reclus contributed an introduction which includes a biographical sketch of his friend. When Reclus accepted the gold medal of the Paris Geographical Society in 1892 he paid warm tribute to Metchnikoff ("more than my friend — my brother in work"). Metchiknoff's daughter succeeded him briefly as secretary to Reclus, and then the work passed to Henri Sensine, professor of literature in the University of Lausanne. Sensine was extremely devoted to Reclus, whom he described as "the purest, finest, and best man I ever knew," but his extravagant praise was sometimes embarrassing. In the prose volume of his *Chrestomathie française du XIXe siècle*, Sensine canonized Reclus in the following tribute:

> The admirable author of the *Nouvelle géographie universelle*, which is one of the most wonderful works of the 19th century, is not only a great scholar in the encyclopedic spirit and the premier geographer of our age, but he is a remarkable man of letters and certainly the greatest French prose writer and one of the best stylists of the 19th century.[37]

Almost twenty years after Reclus's death, when Joseph Ishill was collecting essays for a memorial volume, Paul Reclus described Sensine's contribution in a letter to Max Nettlau:

> Very poor stuff, gushingly admirative, and full of mistakes. I am not to judge if part of it could or could not be made use of; you and Ishill are to decide; but let me know what it shall be, as I have to write to that honest man, and thank him, moreover![38]

If Léon Metchnikoff was Elisée Reclus's "brother," then the young Hungarian geographer Attila de Gerando could have been considered his son. Reclus was very fond of Attila and his sister Antonine and tried to get writing assignments for them. Their father, Auguste, was a Frenchman, and their mother was a Hungarian countess of the Teleki family. The Gerando family lived in Paris during the Second Empire and knew Michelet, Quinet, and Alfred Dumesnil, through whom they were introduced to the Reclus. Elisée became especially close to Attila when they traveled together through Transylvania in July and August 1873. They were also together in Turkey ten years later. Reclus's warmth but also his paternalism were evident in numerous exchanges over the years. For example, in 1877 when Ludovic Drapeyron founded a new journal, *Revue de géographie,* Reclus recommended that both Gerando and Metchnikoff be invited to contribute, even though he himself did not want to be associated with the journal. Reclus wanted Gerando to review the Hungarian section of the *NGU*, but he warned him that he must be sharply critical: "I beg you to be extremely severe. If you cancel or cross things out and cut down the manuscript, you will give me pleasure. . . . The more times you write in the margin 'absurd' or 'wrong', the more you will make me happy."[39]

Reclus's letters to Gerando are very revealing of his ideals and prejudices, although Reclus did not always practice what he preached. In 1874, almost two decades after his last teaching post and two decades before the next, Reclus proffered the following advice to Gerando: "You are perfectly right in not putting geography books in the hands of your students. . . . Books are only for professors: in the hands of students, they generally do more harm than good, they teach truths mixed with errors, but they deprive the child of his intellectual initiative." On another occasion Reclus criticized Gerando's plan to spend only fifteen to twenty-one days in traveling from Venice to Naples: "You risk seeing poorly in seeing too much." This statement is rather ironic, since Reclus himself would probably have covered such a distance in far less time.[40]

In 1879 Reclus, who always retained the bucolic ideal but never seemed to dirty his hands with anything other than ink, praised Gerando for becoming a farmer: "Up to now you have had the misfortune of wandering. As you said, that is bad intellectual and moral hygiene: it is necessary to have a precise goal in life and drive towards it. I also have a little garden, but I scarcely have the time to cultivate it."[41]

The point about having a precise goal was repeated in another letter to Gerando:

Much more than you, I would merit the reproach of our friend Kropotkin, for, although I am a revolutionary by principles, tradition, and solidarity, I concern myself only in a very indirect way with matters of revolution. Apart from some articles, calls, a little oral propaganda, and, from time to time, some marks of solidarity among friends, I do nothing. My life is arranged, not in order to be used directly in the work of social renovation, but in order to be employed in side issues of minimal importance. What I am working at is hardly science, and yet I do not dare to say that I am completely wrong in scribbling each year my volume of banalities more or less passably written. To have a precise work in front of oneself and to do one's best helps to gain respect for one's cause. From that point of view, my work is not completely wasted.[42]

Although Reclus often spoke in this deprecating way about his geographical writing, he did not honestly believe that his books were merely collections of banalities. Surely he would not have spent twenty-two years in writing the *NGU* if he did not believe that it served a useful purpose. In his view, knowledge of peoples and places is necessary for improving understanding. Geography and anarchism are closely related, because the more one understands the world and its inhabitants, the more his prejudices and antagonisms decline, until at last he becomes a true world citizen. If Reclus spoke slightingly of his geographical writings, it was only because he did not want to be known simply as a geographer. He bristled when he was accorded lavish praise by the bourgeois who used his geographical encyclopedia only as a sourcebook of facts, ignoring its subtle message of the brotherhood of man. "Yes, I am a geographer," he said to his friends Roorda van Eysinga and Domela Nieuwenhuis, "but I am above all an anarchist." Actually, both rôles were of equal importance to him; just as his geography was necessary to his anarchism, his anarchism enriched his geography. Reclus cannot be understood if viewed from one side alone.[43]

As "the moral leader of the French exiles in Switzerland," Elisée Reclus was a central figure in the amnesty controversy which raged in France in the late 1870s. To French Republicans the distinguished geographer was a symbol of the need to declare amnesty, and it was with the express purpose of pardoning such men as the Reclus brothers that the Andrieux law offering partial amnesty was passed in 1879. Elisée was swift to reject amnesty until all his fellow communards were fully pardoned. When this occurred, in 1880, Reclus started making frequent trips to France, while retaining his residence in Clarens. In 1882 his father died in Orthez at age 85, and his mother then went to live with her daughter, Zéline Faure, in Sainte-Foy, where she died in 1887 on the eve of her eighty-second birthday. Elisée's daughters, Magali and Jeannie, were married in Paris in 1882 to two of Paul Reclus's (Elie's son) classmates, Paul Régnier and Léon Cuisinier. The double ceremony, performed by Elisée himself, received wide publicity and provoked consternation among the bourgeois. As Reclus had said to his friend Nadar a few days before the wedding: "Neither a priest— representative of a god in whom nobody believes any longer—nor a mayor—spokesman of despicable laws—will be invited to this reunion of the family." He regarded religious or civil marriage as a form of slavery or prostitution, and for it he would substitute "free union," where the couple would be united in the presence of family members and would pledge to each other their love and respect. Reclus was not opposed to the institution of the family—far from it, because he was an unusually devoted family man—but only to the "juridical family," i. e., a family united by law and not by love.[44]

Not only did the wedding of Reclus's daughters offend the sensibilities of bourgeois Frenchmen, but it shocked the Swiss as well, and there were calls for Elisée's expulsion from Switzerland. However, he continued his residence at Clarens until 1890, when he returned to France with his younger daughter. Jeannie's husband, Léon Cuisinier, of whom Elisée had become very fond, died suddenly in 1887, and the young widow brought her three children to Clarens to live with her father. When she returned to France to establish a household in Nanterre in the summer of 1890, Elisée followed soon after. When Ermance sold her house in Clarens and rented a place in Sèvres, Elisée joined her in the autumn of 1891.[45]

When Reclus departed from Switzerland in 1890, he left behind most of the books, maps, and manuscripts that he had accumulated in the preceding eighteen years. In 1893 Charles Perron gave the Geneva Public Library 6,813 maps which Reclus had collected while working on the NGU, and these maps formed the nucleus of the Library's Dépôt des cartes which was established in 1904. Reclus donated 343 volumes and 71 pamphlets to the Geneva Public Library in 1890, and he had previously given them numerous items in the 1880s. In 1890 Charles Knapp of the Neuchâtel Geographical Society asked Ermance

Reclus (with her husband's permission) for the manuscripts of the first fifteen volumes of the *NGU*. The manuscript versions of the remaining volumes were eventually sent to Knapp, and he received the manuscript of volume nineteen in 1897. Today the *NGU* manuscripts (incomplete) reside in the attic of the Institute of Geology of the University of Neuchâtel.[46]

Reclus soon resumed his book-collecting habits, however, because thieves broke into his house in Sèvres and stole some books. Like his father when hungry people took potatoes from his garden, Elisée was not angry with the thieves but with himself for not anticipating their "need." "Now they are going to be of use to others," he said, "Since I did not give them, they did well to take them." This incident well illustrated the anarchist controversy over stealing. To his colleague Jean Grave, who could not condone crime, Reclus said, "We are all thieves and I the foremost, getting ten or twenty times the wages of an honest man." In an article in *La Révolte* in 1889, he said that the real thieves are the speculators and profit-takers, including priests, lawyers, and legislators. "The only difference between the banker and the petty swindler is in the number of operations. The principle is the same."[47]

There was also a controversy among the anarchists over the usefulness of violence—"propaganda by deed"—and Reclus and Kropotkin, as the respected elder statesmen of the order, condoned the acts of desperate individuals although they were themselves incapable of committing violent deeds. These benign gentlemen were badges of respectability for the anarchists, likened by one observer to the bemedaled old soldiers who were hired as doormen by bankers to disguise the nefarious activities within. After 1894, with Reclus in Belgium and Kropotkin in England, French anarchism gave way to syndicalism—whereby anarchist ideals are pursued through militant labor unions—and Max Nettlau expressed the opinion that the continued presence of the leading anarchist theoreticians might have averted the worst features of syndicalism.[48]

In the late 1880s and early 1890s Reclus made the longest trips of his later years. Twice he returned to the United States—in 1889 and 1891—in order to improve the relevant volume of the *NGU,* and in 1893 he traveled to Brazil with his wife. It is odd that he never thought it necessary to travel to distant parts of the Old World while preparing earlier volumes of the *NGU*. He apparently felt the need to return to the Americas (although he did not revisit Louisiana or Colombia) because so many changes had taken place since the 1850s. Reclus delayed the publication of the United States volume of the *NGU* in order to be able to use the results of the 1890 census. When he returned to the United States he saw what he wanted to see, what he expected to see. The life of workers was growing steadily worse, he reported, but the number of millionnaires was increasing. He noted the rural problems of overproduction, absentee landlords, and outsized farms. "Several writers who are behind the times still cite the United States as an example of a land of equality where each disinherited person coming from outside finds a farm and liberty, but pauperism and extreme concentration of wealth are now widespread." He approvingly quoted James Bryce to the effect that "Monarchy, suppressed in the United States, reappears in industry and finance." But however he despised the government, the banks, and other commercial interests (collectively called the "parasitism"), he liked the working people. "The people appear to me to have much cordiality, bonhomie, and egalitarian sense. Daily they are discarding the religious element." Like so many European socialists Reclus misread the political aspirations of the American workers. He expected them to rise up in revolution against their capitalist oppressors, but in reality they were eager to increase their material well-being and even to join the oppressors. In such a fluid society class warfare could not be a serious possibility. In any event, Reclus returned from his American visits with his prejudices intact and his hopes rekindled.[49]

The nineteenth, and last, volume of the *NGU*, *Amazonia and La Plata*, was published in 1894. It ends rather abruptly with a short chapter on the Falkland Islands and South Georgia, and in a brief afterword Reclus explains the lack of conclusion by saying that he planned to write a short book which would not only serve as a concluding volume to the *NGU* but would also highlight its overarching theme of the unity of mankind. This "short book" was to grow into the six-volume *L'Homme et la terre*, of which only the first few chapters were published before Reclus's death in 1905. Just as *La Terre* was the introduction to the *NGU*, *L'Homme et la terre* would be its conclusion. Published over a forty-year period, these three works—in all, twenty-seven volumes—form what Reclus called his "trilogy," but in a sense they form one great work. Reclus planned periodic revisions of the *NGU* volumes and also an annual publication showing new population and production figures, but Hachette was less than eager, and the statistical supplement was dropped after one issue.[50]

With the two South American volumes the great *NGU* came to an end. From the original plan of five or six volumes to be published over a period of four years, this Pantagruel grew to nineteen volumes, the last of which was published twenty-two years after the project was commenced. There is a remarkable degree of unity and coherence in this work. In a review of the final volume, Lucien Gallois praised the entire enterprise, but he specifically criticized its arrangement by countries, which made comparisons and generalizations difficult. He suggested that perhaps Reclus might have chosen a different procedure if he were starting the project in 1894, rather than two decades earlier. Indeed, the Swiss geographer Charles Knapp said that Reclus had told him that if the work were done over it would be reorganized on a different basis. We can only speculate about the alternative arrangements he might have had in mind. Realistically, however, there was probably no other strategy available at that time that would have suited the author, publisher, and reading public than the encyclopedic country-by-country treatment that Reclus actually employed.[51]

As we luxuriate today in our wealth of libraries, it is difficult to appreciate the rôle of the *NGU*, which was a veritable library in itself, in *fin-de-siècle* Europe and America. In 1898 the young American geographer Ellsworth Huntington, while teaching at Euphrates College in Turkey, committed half of the library budget to the purchase of Reclus's encyclopedia. The great work was one of the ornaments of the small library of the Los Angeles State Normal School, the forerunner of the University of California, Los Angeles. For a generation the *NGU* was to serve as the ultimate geographical authority, its eminence confirmed by the plagiarists who pillaged it so freely. "That Providence so often unacknowledged," the historian Lucien Febvre called the *NGU* in 1922, as the aging work was about to be superseded.[52]

The *NGU* is probably the greatest individual writing feat in the history of geography, a feat which is not likely to be matched because there will never be another author or publisher who could mount such an ambitious project and sustain it. It succeeded only because of the longevity and tenacity of its author, and also perhaps because the publishers did not realize what they were getting into. Reclus's mentor, Carl Ritter, had attempted a vast geographical encyclopedia, the *Erdkunde*, but he did not complete it because he got bogged down in his Asian material. Reclus was a more disciplined writer, and he could spend most of his waking hours in reading and writing, unlike Ritter, who had his university classes to tend to. Another *Géographie universelle* was published in the twentieth century on nearly the same scale as Reclus's *NGU*—fifteen volumes (in twenty-three parts) published over a nineteen-year period (1927-1946)—but it represented the combined work of some fifteen geographers. Reviewers of the new work naturally compared it with its predecessor, but Reclus's encyclopedia did not suffer, despite its obsolescence. As P. R. Crowe said in 1928:

> It [NGU] still stands as a remarkable monument to the industry and perseverance of a single man, and most of the criticisms that can be leveled against it may be summed up in the one remark that it is now over forty years old. . . . Yet in some respects the work still has a remarkably modern flavour. This is particularly true of Reclus' emphasis on the importance of the minor elements in topography and his extensive use of parts of large-scale maps as illustrations.

But now the age of the universal geographies—"bibliographic dinosaurs"—must be over. As O. H. K. Spate has said, "He would be a bold geographer who risked his hand now," to which I would add, "He would be a bold publisher who risked his capital now in such a venture."[53]

In 1892 Elisée Reclus, who had not been a teacher since leaving Colombia thirty-five years earlier, received an invitation to become a professor in the Science Faculty of the Free University of Brussels. Actually the title used in the offer was "agrégé," not professor, and there was no indication of the permanence of the position. Reclus readily accepted, but he first wanted to finish the last volume of the *NGU*. Thus he wanted to delay the commencement of his course in comparative geography at the University of Brussels until "the first weeks of the year 1894." The University agreed, and Reclus thereupon set to work winding up his affairs in Paris. If he felt any qualms about leaving France again, this time voluntarily, they were effectively put to rest by a police raid on his apartment in Bourg-la-Reine on New Year's

Day 1894. The raid was part of the police harassment of radicals after an anarchist named Vaillant threw a bomb into the Chamber of Deputies on 9 December 1893. No one was killed by the bomb, but Vaillant was sent to the guillotine, an act which was to result in the assassination of President Sadi Carnot a few months later. The public furor over the terrorist acts caused the harassment of even the gentlest anarchists, such as Elisée and Elie Reclus, and they gladly departed for Brussels in 1894, there to live out their lives. Exile was not voluntary in the case of Elie's son Paul, who fled to Britain in 1894 and lived there for a decade under an assumed name. Elisée probably thought that the Brussels sojourn would only be a rather brief interlude, but instead it was prolonged and became the last chapter of his life.[54]

Chapter Six

Autumn in Brussels

Elisée Reclus's invitation to teach in the Free University of Brussels did not indicate the expected duration of the appointment, but it might have been intended to be open-ended. He perhaps thought of Brussels as just another way-station in his career, but instead it became his happy home and, ultimately, his final resting-place. The Brussels period was the *dénouement* of his long and illustrious career.

The Vaillant Affair quickly reverberated throughout France and Belgium, and Elisée Reclus learned to his dismay that his Brussels hosts had second thoughts about the promises they had made to him in 1892. When public pressure forced the postponement of Reclus's initial lectures in January 1894, the students responded with demonstrations, and Hector Denis, who had been Reclus's most avid supporter, stepped down as Rector of the University. One of the members of the University Council, Dr. Jean Crocq, is reported to have said:

> I disapprove of the social doctrines of M. Reclus . . . I find the social ideas of M. Reclus absurd . . . he attributes to man a perfectability that he will never attain.
>
> But a course in geography has nothing in common with that. M. Reclus is a man of science, the foremost geographer of our era, and the University will be honored to give him asylum . . .
>
> We are a *free* university . . .
>
> Some people find M. Reclus' doctrine dangerous? Others, like me, do not find it dangerous, but absurd.[1]

Actually, the radical dissidents—students and professors—in the Free University had been in ferment for some time, and they seized upon the Reclus controversy as another club with which to beat the administration. Reclus made no overt act to inflame these passions; indeed, he fell ill soon after his arrival in Brussels and was not in proper shape for teaching when he began a series of lectures outside the University in March. In retrospect, it would appear that the controversies could have been resolved without loss of face on either side, but the leftists pulled out and founded their own university—the Université Nouvelle de Bruxelles—which coexisted peacefully side by side with the Free University for twenty years, down to the outbreak of World War I. The New University was formally disbanded in 1919, but its Institut des hautes études still exists.[2]

Reclus's first public appearances in Brussels were in a series of ten lectures on comparative geography beginning 2 March 1894, delivered before turn-away crowds of five hundred enthusiastic listeners in the meeting-hall of Les Amis philanthropes. Doubtless many people came to the first lecture more out of curiosity because of the speaker's notoriety than out of interest in geography, and some political reactionaries might have been present to see if his radical views would intrude, but his calm and gentle manner won them over, and everyone seemed to regard the exposure to Reclus as an uplifting experience. It is difficult to imagine how he could have captivated such large audiences, for his voice was not powerful and his manner was not charismatic, but his saintly aura was somehow felt even in the outermost reaches of the hall.[3]

Whether in these earliest lectures or later in the year, Reclus made a true convert of a well-to-do Brussels widow, Mme. Florence de Brouckère, and she became his close companion for the rest of his life; indeed, Reclus spent long periods at her country home at Thourout, and it was there that he died in 1905. It is difficult to determine the nature of Reclus's relations with Mme. de Brouckère, but it would appear that they were only close friends. He had found in her "a new Fanny," i.e., a woman much like his second wife. Ermance retreated into the background during the Brussels period.[4]

The New University took shape during the summer and autumn of 1894, and Elisée Reclus taught his first class on October 27. The course which he offered in the academic year 1894-1895, "Comparative Geography in Time and Space," was a series of lectures on Asian culture history with an environmentalistic flavor. These lectures, as well as those of subsequent years, which continued to explore Asian themes, were employed in the composition of his last major work, *L'Homme et la terre* (six volumes, 1905-1908).[5]

Initially the professors were mostly leftists of various hues. None was paid for teaching; they had to have other means of support. Elisée Reclus was never paid for teaching in the New University, but after 1898, when he established a geographical institute and developed a publishing program, he was able to provide some remuneration for cartographers and lecturers in the institute.

The New University was established on principles which were basically the same as those of the Free University but with a pronounced socialist bias and with avowed adherence to Comtian positivism. Following Auguste Comte's dictum that nothing can ever be properly elucidated except through its history, historical lectures were a necessary complement to the dogmatic or factual treatment of a subject. That is perhaps why Reclus was recorded as offering both Geography and the History of Geography. In fact, he achieved a blend of geographical and historical approaches in his lectures on the comparative geography of Asia. He had never been terribly keen on Comtian positivism, but he was willing to go along with De Greef and his other Belgian colleagues.[6]

Reclus was happy in teaching at the New University, although it could not develop into a power base such as a professor would enjoy at the University of Paris. The diplomas of the New University were not recognized by the State, and so the Faculties of Science and Medicine—necessary for a true university—could not develop. For that reason also, few Belgian students were attracted to the New University, but foreigners appeared in some numbers, especially from Eastern Europe. As an egalitarian, Reclus abhorred all diplomas, even the worthless ones granted by the New University. Although he had high hopes for the University because of its relative freedom from the State and from political parties, he thought that the students, like university students everywhere, were a privileged class which had an unfair advantage in the battle of life. Although the University's motto was "Let Us Make Men," Reclus could see that they would inevitably make exploiters. He was much more interested in extension courses, which reached the public but did not create "Bachelors" or "Doctors." As he said to his friend Roorda van Eysinga in 1895, "Our university is an institution like any other—therefore bad—but for the moment it represents the struggle."[7]

Just as Reclus felt that books and even maps were often impediments to learning, he thought that classrooms did not provide the proper atmosphere for students. The real classroom is out-of-doors. Direct observation of Nature through travel would provide a better education than the schools. Travel had been democratized by the lowering of costs through improved transport technology and was now within the reach of all economic classes. Reclus optimistically believed that poor folk benefit from travel more than the wealthy because they are better able to see and appreciate other lands and peoples. However, nature study and travel have to be disciplined or directed; haphazard travel can produce boredom and even kill one's interest in natural beauty. Reclus was aghast at the ignorance and insensitivity of American tourists, who were already proverbial for their whirlwind tours of the Old World, during which they managed to add only misinformation to their already inadequate and incoherent educations. "Everything is strangely mixed in their memories—the balls of Paris, the review of the guards at Potsdam, the two visits to the pope and sultan, the climbing of the pyramids, and kneeling before the Holy Sepulchre." He recoiled at the bored young man who announced his forthcoming visit to Mont Blanc in a resigned tone, "It is however necessary that I go see that junkpile." Reclus was no less critical of those who served the tourists, as when he noted that Switzerland had been defiled by "insolent architects, paid by obscene innkeepers." In that country, he said, " 'The exploitation of the stranger' is one of the principal industries. . . . One seeks to appropriate the beautiful sites to make others pay dearly for the view of them, and more than one cascade is made ugly by the frightful fences which protect it from the glances of the poor."[8]

Reclus was not a practiced orator, and he had no real classroom teaching experience before he went to Brussels. His teaching in England, Louisiana, and Colombia in the 1850s and in prison in Brest in 1871 was really only tutoring, teaching a single student or a handful of students. He had read papers before the Paris Geographical Society and had given some public lectures in Switzerland, but he did not become a classroom teacher until he was sixty-four years old. I have not been able to determine whether he read his lectures in March 1894 or whether he spoke extemporaneously, but his friend Edmond Picard later recorded that at an early meeting (October 1894?) of the professors of the New University it was decided that all lectures would be verbal—*i.e.*, extemporaneous rather than read—and Reclus startled the group by saying: "Gentlemen, I myself cannot lecture in that fashion; I have always read my lessons." The other professors were a little embarrassed, and they quickly agreed that an exception would be made for him, but he made up his mind that he would conform to the new system, and he never again read his lectures. He spoke authoritatively, without hesitation, and with such great warmth and verve that his manner utterly concealed his inner fears and his extensive preparation.[9]

Although Reclus did not develop disciples in the conventional sense, he nevertheless won a large following among his Belgian students and younger associates. He developed a particularly close friendship with Jacques Dwelshauwers, who later adopted the pseudonym of "Mesnil." Jacques was the younger brother of Georges Dwelshauwers, who had provided a *cause célèbre* that had plunged the Free University into controversy before Elisée Reclus came onto the scene. Jacques and his friend Clara Koettlitz, who were joined in "free union" (a marriage like Reclus's) in 1897, were thoroughly devoted to the old anarchist, and he shared with them some intimate thoughts that were kept hidden from many older acquaintances and even family members. Jacques Mesnil became a militant anarchist and art historian, specializing in Renaissance painting, especially the works of Botticelli. It is interesting that toward the end of his life Mesnil made his living in much the same way that Reclus began—by working on travel guides, in Mesnil's case the *Guides bleu,* the post-World-War-I successors to the *Guides Joanne.* It is curious, but not surprising, that Mesnil as an older man deplored the democratization of travel brought about by the introduction of the automobile and looked nostalgically back to the bourgeois Baedeker style of the nineteenth century. Many of Reclus's younger associates thought that Mesnil would produce the definitive Reclus biography, but he died suddenly at the time of the German invasion in 1940, by which time he had written a fifty-page typescript which recounted Reclus's life only up to 1868.[10]

Not all of the younger people who swam into Reclus's ken were willing to accept discipleship, however. Mesnil expressed contempt for those of Reclus's visitors whose motives were not the highest:

> A great number of young men came to visit him with inconceivable nonchalance and on the pretext that he was anarchist, like themselves, and a "comrade" like all others, treated him without any regard, and gave proof of a vulgarity and indelicacy which saddened him. That is why, in his last years, he did not feel a greater sympathy with those so-called anarchists whose entire anarchy consists in licentiousness of spirit, grossness of manner, and an idiotic equalitarianism which is very convenient to their petty vanity which prevents them from recognizing any superiority in another.[11]

Even those who approached him more respectfully did not remain long in the ranks of the faithful. Mesnil complained of their inconstancy, but he well understood it.

> The majority of the young people came to anarchy for sentimental reasons: out of generosity in their hearts, out of youthful enthusiasm, out of discontent with the present social situation, out of admiration for a man like Reclus, or by impulse. Only a few of them were anarchists by temperament. . . . In the others it would have been necessary to constantly reinforce anarchy. Reclus had encouraged and counseled them, but then, all he could was let them fly with their own wings; anarchy is not a doctrine to be taught, nor a religion that has answers for everybody.[12]

A person who admired Reclus greatly but did not come under his spell was Emile Vandervelde, the future Socialist politician. Vandervelde knew both Elie and Elisée and, not surprisingly, preferred the former for his sweetness and for the constancy of his libertarianism. He found the younger brother charming but dogmatic. Elisée's abhorrence of politics could not be accepted by the young socialist, who took the practical view that only those in office can implement policy. Vandervelde described their differences in the following portrait of Reclus:

> An incomparable conversationalist, the man certainly had a magnetism of irresistible charm. It was a delight to hear him affirm, in the manner of Rousseau, at the end of each of his lessons, his unalterable

faith in human goodness, free of all the impurities and blemishes of a social state based on force.

But on closer contact, one encountered with him, underneath the extreme agreeability of his words, something that I had later found again with Kropotkin: a sort of closedmindedness or even hostility to all socialist thought which was not identical to his own, which took account of realities that he denied, or which was unwilling to admit that the state could not be, or become, something other than the state of force, at the exclusive service of the proprietary classes.[13]

Reclus's light teaching duties enabled him to turn his attention to some elaborate cartographic schemes which he had had in the back of his mind for several decades. The idea of building a giant globe apparently came to him when he viewed Wyld's globe in London in 1852.[14] Reclus was to retain a vivid impression of Wyld's globe in his mind for more than forty years until he was in a position to suggest an even larger one. After settling in Brussels in 1894, he revived his old dream of building a giant relief globe which would not only be a scientific showpiece but a beacon to mankind. The proposed Paris Exposition of 1900 would provide the opportunity for constructing the globe. Reclus's name had been invoked as a source of inspiration for the Villard-Cotard globe at the Paris Exposition of 1889, but he had played no real rôle in its planning or construction. The great marvel of the 1889 Exposition was, of course, the Eiffel Tower, which became such a prodigious revenue producer that it was not demolished, as the nearby globe was in 1891. Subsequently, planners of grandiose schemes for other expositions hoped to imitate the success of the Eiffel Tower, the "clou" or hallmark of the 1889 fair. Reclus hoped that his globe would be the clou of the 1900 exposition. (The idea that every world's fair should have a clou has carried over to this day, and some of the clous have had a geographical character—e.g., New York's "Unisphere" of 1964.)

Reclus first planned a relief globe at the scale of 1:100,000 (diameter about 420 feet, ten times the diameter of the Villard-Cotard globe.) The plan was set forth in a pamphlet published in 1895 entitled "Projet de construction d'un Globe terrestre à l'échelle du Cent-millième." Elie's son, Paul Reclus, who had fled from France during the police crack-down on anarchists the previous year and was living in

England under the pseudonym of Georges Guyou, provided an appendix with the technical details. Elisée's son-in-law Paul Régnier, an architect and one of Paul's fellow students in the Ecole Centrale fifteen years before, also provided technical advice. In 1896 Paul moved to Edinburgh to aid in several cartographic projects at Patrick Geddes' Outlook Tower, and so he was able to serve as a liaison between Geddes and Elisée Reclus.

Reclus's "Projet" called for a "map of the heavens" on the upper part of the inside of the globe. When word of this reached the Paris architect Albert Galeron, he wrote anxiously to Reclus to express the hope that this map would not conflict with the "Cosmorama," or celestial sphere, that Galeron had proposed to the 1900 Exposition Commission in July 1895. Galeron subsequently drew up plans without Reclus's knowledge to combine the two globes—his own Cosmorama and Reclus's Great Globe—and Reclus, after testily asserting his independence when opportunistic individuals urged a fusion of the plans, finally acceded to the merger in the spring of 1897, probably in the belief that this now offered the best chance for success. He had earlier accepted an alliance with Mme. A. Bressac, "a courageous, enterprising, and resourceful woman, overflowing with enthusiasm and faith," whose Diorama was to consist of "photographs . . . of all the principal monuments of the earth." E. J. Marey's "Panorama chronographique du développement de la civilisation," a chronological panorama of human history which was to be anywhere from 120 to 150 meters long, was willingly added to his own scheme in 1897. It would appear that Reclus, "qui n'a rien de *businessman*," could see that his plan would stand a better chance if it were merged with others. He tried, however, to maintain control over the combined ventures so that the scientific aspects would not be subordinated to the commercial. Reclus's agent in charge of technical and commercial matters was Lt. E.–A.–L. Hourst, who had earlier engaged in exploration in the Niger River country. Hourst could see the usefulness of the Globe and its attendant geographical exhibits in promoting colonial ventures, a view which Reclus himself would not share.

At first, in 1895, Elisée Reclus proposed that the Great Globe be built at the scale of 1:100,000 with a shell of 1:80,000 around it. Financial and technical considerations soon forced him to build a smaller globe, and scales of 1:200,000, 1:250,000, 1:320,000,

1:400,000, 1:500,000, and 1:1,000,000 were considered. In late 1897, when Reclus was planning a globe at the scale of 1:500,000, he noted that it would be insufficient to show even rather larger relief features. What is needed, he said, is a sphere of 1:50,000, where even little hills, such as the Trocadéro in Paris, would be shown. He admitted that such a globe would be rather large (diameter about 840 feet!), but he added, laughing, "Why wouldn't it work out? Nebuchadnezzar dreamed of building a monument which reached the heavens. We would not go so high."

The Globe plan naturally attracted criticism, not for the ultraist tendencies of its author, but for the excessive cost of the venture, estimated at anywhere from twenty to fifty million francs. "Certainly the sum is great," said Reclus, "but it shouldn't frighten us, for it represents a necessary work which humanity can't do without in order to arrive at a perfect knowledge of its domain, and we know, alas! in how many futilities and even crimes our human resources are wasted." Even those who were sympathetic to the scheme questioned whether the money might not be put to better use. Nor did the Globe escape artistic criticism. When Reclus thought of erecting his Globe in the Trocadéro gardens, one critic thought that the "iron egg" would not be as esthetically pleasing as the flowerbeds it would displace.[15]

Reclus had planned a relief globe with winding staircases or some sort of tramway around the outside so that viewers could observe the earth's surface closely. The surface would be made up of removeable panels so that it could be kept up-to-date as more information about the physical features of the earth became known. One of the stated advantages of the globe was that it would facilitate the making of maps, at any scale and of any part of the world, by photography. However, Alfred Russel Wallace, who had long admired Reclus and who, indeed, had petitioned for the softening of his sentence in 1871, wrote a thoughtful critique of the Great Globe scheme in 1896 in which he demonstrated that accurate maps could be produced by photography only if small sections of the convex surface were photographed. Wallace urged a return to Wyld's idea of putting the earth's surface on the inside, concave surface of the globe. The viewer could then comprehend much more of the earth's surface than he could if he were viewing it from the outside, and, furthermore, photographs of the concave surface would produce accurate maps. Although Reclus greatly admired Wallace, he nevertheless adhered to his plan of modelling the relief on the outside because he wanted to remain faithful to Nature.

Early in 1897 it was hoped that the Globe could be sited at the Place de l'Alma in Paris, and Galeron drew a pleasing sketch of what it might look like, but it was feared that it might not be possible to win a long-term concession there, and in November 1897 Reclus's interest shifted to the Place du Trocadéro. He then petitioned the Municipal Council of Paris for a concession on the Trocadéro site, and on 31 December 1897 the Council unanimously voted to give him three months to obtain the necessary financial backing. Of particular interest is A. Thuillier's favorable report to the Council in December 1897. Thuillier stressed the scientific and educational value of the enterprise, "which would bring the greatest honor to French science." "Formerly one could have described the Frenchman as being recognizable by his small moustache and his absolute ignorance of geography, but, thanks to improved instruction in the schools, the progress of exploration, and now this globe project, this ignorance is being dispelled," said Thuillier.

In the first quarter of 1898 Elisée Reclus allied himself with Charles Bivort in an attempt to establish a corporation capitalized at some three to four million francs for the purpose of constructing the Globe. When they failed to meet the deadline at the end of March, there was still hope that the Municipal Council would grant an extension, but Reclus did not fight for more time, because he was then setting up an institute and corporation in Brussels in order to finance not only the Great Globe but his other geographical ventures as well. He still hoped to place a Globe in Paris, with or without connection with the 1900 Exposition.

In the 1890s six institutes were created within the Université Nouvelle. The oldest and longest-lived—indeed, it survived the dissolution of the University in 1919 and continues to this day—was the Institut des hautes études. Some of the other institutes were highly specialized, such as the Institut des fermentations, which taught the arts of brewing and distilling. Elisée Reclus founded the Institut géographique on 18 March 1898 and then established the Société anonyme d'études et d'éditions géographiques Elisée Reclus, which was designed to support the Institut and its auxiliary projects. (N.

Drawing by Albert Galeron of Reclus's proposed Great Globe at the Place de l'Alma, Paris, 17 March 1897. Reproduced through the courtesy of the Bibliothèque Nationale.

A drawing of Reclus's proposed Great Globe ("oeuf en fer") in *L'Illustration*, vol. 111, no. 2871 (5 March 1898), 183.

[109]

B.—A "société anonyme" is a corporation, not a scholarly society.) It was hoped that through the sale of stock the Société would be able to support the Great Globe and numerous other cartographic and publishing ventures, including the publication of Reclus's *L'Homme et la terre*. In 1899 various geographical publications were proposed, including *L'Actualité géographique*, a daily (?) publication of a map or maps of places currently in the news, but these plans never came to fruition. The Société had only a brief and tempestuous existence (1898-1904), but although it cast its founder upon the shoals in 1900, it nevertheless had enabled him to build up a small cadre of geographers and cartographers who were able to carry out a few of the projects despite the financial stringencies. The crisis stage was reached in 1899-1900, and Elisée was forced to borrow money from his son-in-law, Félix Ostroga, who had married Jeannie Cuisinier, and from Henri Sensine. Friends helped Elisée avert bankruptcy, but he was left in rather straitened circumstances. Probably the Société anonyme was doomed from the start, because "all responsibility rested," as Paul Reclus said, "on the shoulders of Elisée and Elie, the most ignorant people the world ever produced on legal and financial matters."[16]

Despite its shaky foundation the Institut made possible the expansion of the teaching of geography and cognate disciplines in the New University. The University did not pay instructors, but Elisée was able to find a small amount of money to support cartographic work and some teaching. He even tried unsuccessfully to attract his old cartographer from Geneva, Charles Perron. Emile Patesson of Brussels became allied with Reclus, and he was not only an excellent cartographer, but he provided moral and even physical support to Reclus in the period of crisis. Paul Reclus said that at the height of the crisis Patesson stood in the doorway, "in the position of a boxer," denying entrance to Reclus's creditors. Patesson became a professor in the Institut géographique, and he had his own cartographic establishment in Uccle (Brussels) up to the beginning of World War II.[17]

Another recruit for the Institut was a versatile young naturalist from Bordeaux, George Engerrand. Reclus had apparently heard of him through some Gascon connections, and he invited Engerrand to come to Brussels to teach biology, physical geography, and anthropology in the Institut, with a guaranteed salary of at least a hundred francs a month. It is very interesting to see how Engerrand, who was to enjoy a long career as a professor of anthropology in the University of Texas, got his start. Writing to Engerrand in June 1898 on behalf of her brother, Louise Dumesnil, who had come to Brussels to serve as Elisée's secretary, said:

> You have studied biology, but my brother thinks that even though you would be less well up on anthropology nothing would prevent you from studying it and then designing courses. . . . Here you will manage in such a way . . . that your parents, after having barely tolerated you, will end up being proud of you. Evidently you could thus avoid military service.[18]

Reclus's subsequent communications with Engerrand are mostly written in a formal and paternalistic tone. Using his granddaughter Magali Cuisinier as an amanuensis, Reclus advised Engerrand not to come to Brussels during the long summer vacation: "If you cannot remain in Bordeaux you might spend the time in the country with some good peasant helping him dig up his land." In a subsequent letter Reclus solemnly averred that "each day is a battle and . . . it is necessary to conquer each minute of life, if one does not allow himself to drift like a derelict." When he sent Engerrand some money in late September in order to travel to Brussels, Reclus said that Engerrand certainly had the necessary knowledge to teach in the Institut but that he would also have the requisite moral qualities: naïveté, simplicity, modesty, cordiality, and solidarity. Engerrand taught in the Institut until 1907, spent the next decade in Mexico, and then taught anthropology for forty years in the University of Texas, retained long after the normal age of retirement because of the excellence of his teaching. When he died in Mexico City on 2 September 1961 he was engaged in writing the biography of his old patron—Elisée Reclus—but unfortunately he had only begun to write the introductory chapter. The heading of the first page of the manuscript—"A Life without Blemish"—sets the tone for what would have been a highly adulatory biography.

Apart from maps, the Institut géographique published a monograph series, consisting of ten rather brief studies, beginning with Siemiradzki's paper on New Poland (Brazil) in 1899 and ending with Sacré's study of Esperanto in 1905. Several of the papers were transla-

[111]

tions of articles originally published in other languages, especially English. The only monograph longer than 37 pages was Kropotkin's "Orographie de la Sibérie avec un aperçu de l'orographie de l'Asie," a monograph of 119 pages published in 1904.[19]

After Reclus's death in 1905, the Institut was directed by nephew Paul until 1914, when the New University was closed for the duration of the war. The Institut's library, which was said to number some 40,000 items in 1914, was sold after the war to a Japanese man, a Mr. Ishimoto, who wanted to establish an Institut de Géographie Elisée Reclus in Tokyo. The collection was shipped to Japan in 1923 and managed to reach the docks of Yokohama just in time to be destroyed by the great earthquake and tidal wave.[20]

Elisée Reclus's greatest work of the Brussels period was the six-volume L'Homme et la terre, which completed his trilogy which began with La Terre, but unfortunately only a few fascicules of the first volume of L'Homme were published before his death. In this work, which amounted to 4,500 pages of manuscript by the time of its completion in 1904, Reclus used his lectures on comparative geography in the New University and also his other publications of the last decade of his life. It is interesting to see how the titles of Reclus's trilogy mirror the evolution of his geographical thought—from La Terre, in which man appears in the final chapters almost as an afterthought, through La Terre et les hommes (subtitle of Nouvelle géographie universelle), to L'Homme et la terre, where man assumes dominance. La Terre had been largely a descriptive physical geography, the Nouvelle géographie universelle was a world regional geography, and L'Homme et la terre was intended to be a social geography—the capstone of his life work. He described his intentions in the first lines of the preface to volume one:

> Several years ago, after having written the last lines of a long work, the Nouvelle Géographie universelle, I expressed the wish to be able one day to study Man in the succession of ages as I had observed him in different countries of the globe and to establish the sociological conclusions to which I had been led. I drew up the plan of a new book where would be exposed the conditions of soil, climate, and all the environment in which the events of history are accomplished, where would be shown the accord between men and the earth, and where the activities of people would be explained, from cause to

effect, by their harmony with the evolution of the planet. . . . I knew in advance that no research would make me discover the law of human progress . . . but we can at least . . . recognize the intimate connection which attaches the succession of human facts to the action of terrestrial forces.[21]

Actually, it took Reclus a long time to settle on the title for the third part of the trilogy. By 1872 he had already referred to it as *L'Homme,* but *terre* was not added until the first *livraisons* were published more than three decades later. Some proposed titles were *L'Homme à travers les contrées et les âges, L'Homme, Essai de géographie sociale,* and, in the event that it was published first in English, *Comparative Geography and History.* The term "social geography" is especially interesting because of its heavy use in the 1970s. Reclus has been credited with being the first writer to use it, in the first volume of *L'Homme et la terre* (1905), but he actually used the term ten years earlier, and my cursory search has turned up its use by the Le Playist sociologist Paul de Rousiers in 1884. Reclus did not offer a definition of social geography. He equated it with historical geography and said that it concerned three orders of facts: the class struggle, the search for equilibrium, and the sovereign decision of the individual. In an essay on Reclus published in 1961, the Soviet scholar G. S. Tikhomirov specifically criticized this conception of social geography, because Marxists are interested only in the class struggle and not in equilibrium or individualism. Tikhomirov found much to praise in Reclus's work, even though he thought that it suffered from the faults that are inherent in the anarchist viewpoint.[22]

After the completion of the last volume of the *NGU* in 1894, Hachette expected to publish revisions, statistical supplements, and also *L'Homme,* which the author had regarded as the conclusion of his great geographical encyclopedia. In October 1894 Hachette proposed a six-volume *Géographie populaire* — one volume on France and five on the rest of the world—because sales of the *NGU* had dropped, and they were unwilling to consider reprinting all nineteen volumes. However, the idea of the *Géographie populaire* was shelved two months later. The statistical supplement was dropped after one issue, in 1894, the South Africa and China sections of the *NGU* were edited by Onésime Reclus and published by Hachette in 1901 and 1902, respec-

[113]

tively, in order to capitalize on public interest in the Boer War and Boxer Rebellion, but *L'Homme* was finally done by another publisher, despite Hachette's avowal of interest in publishing the work as late as 1904.[23]

The rift between Hachette and Reclus seems to have resulted mostly from the latter's increasing crankiness. He imagined that Hachette wanted to censor his works in order to placate their bourgeois readership, but, in fact, Emile Templier had allowed Reclus free rein right from the start of the *NGU*. After making stylistic suggestions for the first volume, Templier did not read Reclus's material before it went to the printer. Templier died in 1891, and his son-in-law, René Desclosières, had less success, despite his considerable diplomacy, in dealing with the aging author. Sales of the last volumes of the *NGU* slumped badly, and Reclus's account with Hachette was in arrears by almost 32,000 francs in September 1894. The publishers offered to cancel this debt, but Reclus asked them if they would make him an advance on the unsold livraisons and also buy his library. Desclosières' reply of 7 December 1894 shows the continuing affection of the Hachette house for Reclus:

> Each year we will give you an account of your royalties on the copies which are sold; but as we fear that this will be only a small number, we offer to give you for ten years, beginning 1 November 1894, a monthly sum of 833 F 33, assuring you of 10,000 F a year.
>
> As for the library which you have offered to us, we want you to retain the free disposition of it . . . so that you can always dispose of it as you wish.[24]

Reclus quickly and gratefully accepted this offer, and he thanked Desclosières for his "generous largesse." The income of 10,000 francs a year was paid by Hachette right up to 1905, the year of Reclus's death, even though the sales of the *NGU* volumes were very slow. In 1900 there still remained in stock more than four million livraisons, or about 70,000 volumes. Finally, the publishers had to shred two-thirds of them (2,700,000 livraisons).[25]

Despite Hachette's generous gestures, Reclus began to cool in his relations with them. In June 1895 he wrote the following letter to Desclosières:

You doubtless remember that you agreed to publish my work in preparation: L'Homme, Géographie sociale, but only on the condition that my conclusions would not be of such a nature as to offend the usual readers of works published by your house. That decision not permitting me to work with security towards the continuation of what I call the end of my "Trilogy," I waited for the opportunity to get a firm response from another publisher. I did not think for a minute that that other publisher could be a Frenchman, because of a natural feeling of respect for my hosts of forty years, and I got in touch with . . . one of the London publishing houses. A first attempt failed because of a misunderstanding, but a second succeeded, and, for several months already, the contract has been in preparation. I must sign it today.[26]

Arnold was the English publisher interested in L'Homme, but in 1903, when it came right down to publishing the work, Arnold backed off from it, particularly because Reclus had insisted on a large number of maps, which would have been too costly to print. Hachette thereupon renewed its offer to publish L'Homme, but then retracted it, according to Reclus, "because members of the army, the law, and the clergy would refuse to buy." Onésime then put Elisée in touch with the Librairie Universelle, the ultimate publishers of the work.[27]

Actually, there is no evidence to show that Reclus's quarrel with Hachette was anything more than one-sided. He seemed to be spoiling for a fight, and the old anarchist was probably relieved when he cut off his forty-year association with the large publishing house.

His new connection, with the Librairie Universelle, did not prove satisfying to him either. He soon complained about the impersonality of the publishing and printing processes:

I work at the book without conviction and thus without pleasure: editor, publisher, printer, illustrator, and the proofreader do not know each other: the work is without unity and I am becoming nearly hostile to it. . . .

My book does not give me pleasure. . . . the work is practically produced industrially and I do not count. Like the Florentines in their good times, I should have cut the letters for the printing of the book myself.[28]

Actually, Reclus was very fortunate in his illustrators. Emile Patesson did many of the maps, performing the same rôle for L'Homme as Charles Perron did for the *NGU*. The striking drawings for the beginnings and ends of the chapters were the work of François Kupka (1871-1957), a Czech artist who settled in Paris in 1894. Kupka took his assignment to heart; he studied sociology and ethnography in order to make realistic illustrations for *L'Homme*. These drawings gained him a considerable reputation and numerous commissions for book illustration. His enduring fame, however, came a few years later whan he changed his style and became a pioneer of abstract art. Kupka had come to Reclus's notice because he fancied himself to be an anarchist, but he was only one in a very loose sense. He was an extreme individualist, a libertarian, but not a true anarchist of the sort with whom Reclus could be happy. As a puritan ascetic Reclus could not be comfortable among artists.[29]

The first fascicule of *L'Homme* was issued 15 April 1905, less than three months before Reclus died, and he was able to make some slight corrections in the first 300 pages of his manuscript. He begged his collaborators to recast completely several chapters with which he was not satisfied, but his wishes were not fully carried out, and the published text followed the original plan rather closely. Nephew Paul took great pains to make the final version conform to Elisée's style, while disguising his own substantial contribution. The first volume was published in October 1905, and the last appeared three years later.[30]

Of all parts of Reclus's trilogy, *L'Homme et la terre* is today the most valuable for the geographer, but, paradoxically, it is the least well known, probably because it was not handled by Hachette's successful marketing organization and because it was not translated into English or other languages. *La Terre* went through several editions but was woefully obsolete by the 1890s; only its coda on Man endures. The first

---

Three illustrations by François Kupka for Reclus's *L'Homme et la terre* (The figures at the beginning of each chapter were drawn by Kupka and engraved by Deloche.):

A. Chapter heading of Chapter 3, "Latins et Germains," of Volume 5 (p. 377). Kupka's wife was the model for the figure at the left representing France.

B. Drawing at the end of Chap. 9, "Iles et rivages helléniques," of Vol. 2 (p. 426)

C. Chapter heading of Chapter 12, "L'Inde," of Vol. 3 (p. 101)

volumes of the *NGU* were composed two decades before their youngest siblings, but the work retains a slight usefulness today, just as any antiquated encyclopedia would. *L'Homme*, however, compels our interest, even though it is very definitely a period piece. Reclus called it a work in social or historical geography, although the noun and qualifiers should be reversed. Emphasis is not so much on human geography—location, distribution, and spatial interaction—as it is on events and institutions. As the volumes progress from ancient times to modern, the geographical content declines, and the discussion of contemporary institutions is quite journalistic in its approach. *L'Homme* is a universal history worthy of Arnold Toynbee or any other acknowledged master of that genre. It is redolent of Reclus's anarchism and also of Social Darwinism. It is of interest today for the anarchist's view of history, with its emphasis on those elements that are of perennial concern to the engagés—social ills and their solutions. *L'Homme* strikes a responsive chord with anyone who favors decentralization and the dismantling of bureaucracies. It is not surprising, then, that Karl Marx was barely mentioned in Reclus's works and then only as a misguided proponent of centralization and state socialism. "Ni Dieu ni maître," Reclus might have said, anticipating the modern slogan. His alternatives to capitalism and Marxian state socialism would have been anarcho-communism and federalism. The current slogan "Small Is Beautiful" would have enjoyed eager adoption by Elisée Reclus.[31]

One interesting passage in *L'Homme* that has a modern ring is a comment on the ideal spacing of cities, a matter that Reclus had treated in a paper, "The Evolution of Cities," published in the *Contemporary Review* in 1895. Reclus observed that, given a fair degree of physical uniformity, "the larger towns [are] . . . rhythmically spaced . . . each possessing its planetary system of smaller towns." Although Patrick Geddes has credited Reclus with originality in making these observations, it would appear that they derived from the works of J. G. Kohl and other earlier writers. Reclus himself cited an obscure predecessor who has so far proved to be rather elusive:

> A little pamphlet [*Le Gerotype* in a footnote] written in 1850, or thereabouts, by Gobert [*sic*], an ingenious man and an inventor, living as a refugee in London, drew attention to the astonishing

regularity of the distribution of the large towns in France before mining and other industrial operations came in to upset the natural balance of the population.[32]

Another item that Geddes might have borrowed from Reclus is the observation that cities have a tendency to grow westward, so that, in a significantly large number of cases, the highest priced residences are in the western parts (the West End) of the cities. As Reclus said, "The rich, the idle, and the artist are more apt to enjoy the beauties of the twilight than those of the dawn."[33]

In *L'Homme* Reclus reiterated his concerns about the unequal distribution of wealth and underscored the importance of the study of geography in making an inventory of the world's resources and suggesting a plan for their equitable spread.

> Mankind has not yet made an inventory of its riches and decided in what manner it ought to distribute them so that they might be best divided for the glory, the profit, and the health of all people. Science has not yet stepped in to determine which parts of the earth's surface should remain in their natural condition and which should be used otherwise, either for the production of food or for other elements of the public welfare. How can one ask society to apply the teachings of statistics when she has declared herself helpless against the proprietor or the individual who has the "right to use and abuse." . . .
>
> The first thing to do would be to introduce order and reliability into distribution; it would consist of sending and dividing the various products—grains, vegetables, and fruits—with as much method as in bringing everyone, early in the day, his letters and newspapers. . . .
>
> At present, in every country, the number of commercial transactions is taken as an index of prosperity. The contrary point of view would be more logical: the better the land is utilized by its inhabitants, the less becomes the necessity of moving goods over great distances; the more sensible the work of their factories, the less becomes the exchange of products.[34]

Reclus would argue that the basic decisions about production and consumption should not be made by bureaucrats but by communities or associations of free laborers who have transcended avarice and the lust for power. The goal of "free production and equitable distribution" is attainable, and Reclus was convinced, in his historical survey, that mankind was definitely making progress toward that happy state. His undying optimism suffuses the six volumes of *L'Homme* and gives them their unifying thread.[35]

Just as Elisée was completing the manuscript of *L'Homme* that had taken a decade to write, his brother Elie, his inseparable companion, died. Actually, the devoted brothers were only separated for a year and a half, because Elisée was interred in the same modest grave. It might seem odd that they were buried together and not with their wives, but in this case the fraternal bond was closer than the matrimonial one, even though they were loving husbands and fathers. Elisée took his brother's death hard—harder, in fact, than Elie's wife and son Paul did. Elisée published a thirty-two-page eulogy of Elie that emphasizes the early years and is in part autobiographical because it reveals so much about Elisée himself.[36]

It is truly remarkable how the brothers' basic personalities remained essentially unchanged throughout their long lives. Elie was always described as the more thoughtful and reserved one, with a basically pessimistic nature, whereas Elisée was characterized as enthusiastic and optimistic. The differences between them were especially well limned in an essay by their nephew Elie Faure, the well known art critic. Faure called his uncles "Le Phare et l'astre" ("The Beacon and the Star"):

> Elisée was a beacon in the distance. One knows that a beacon can go out and that afterward dawn will break. Elie was a natural star that rises and sets. Think of what horror one would feel at no longer seeing it one morning.[37]

Faure was especially fond of uncle Elie, but he had a very high regard for Elisée. He saw profound differences between the two, even though they had been raised together.

> One [Elie] came manifestly from Montaigne . . . the other manifestly from Rousseau. The one penetrated with pagan culture, accepting that man was what he is . . . . The other, Christian at bottom, affirming that man is not what he appears to be . . . . Both so pure that the pessimist always kept the freshness of sentiment and the laughter of youth and that nothing, not even unhappy experiences, could make the optimist believe that every revolutionary movement was not the seventh fanfare of the trumpets of Jericho.[38]

Faure recalled that when he visited the Paris Exposition of 1900 in Elisée's company, he was assured by his uncle that anarchy would reign over the earth by 1906 at the latest. On this same occasion Elisée allegedly said to his nephew: "I am an anarchist in the individual order, socialist in the social order, and republican in the political order," but in another essay Faure put these words into his own mouth. This particularly Gallic expression would seem to apply more to Faure than to his uncle, but it is interesting nevertheless.[39]

Paul Reclus, Elie's son, returned from exile only a few months before his father's death. He moved his family from Peebles, Scotland, to Brussels and took up a position in his uncle's Institut. The failing Elisée saw in Paul the perfect replacement. As he said to Nadar, "My nephew . . . will complete me, enlarge me, and extend me." Highly intelligent and dedicated, although modest to a fault, Paul completed all his uncle's writing projects, including the monumental L'Homme et la terre, and directed the Institut down to the outbreak of World War I.[40]

One of Elisée Reclus's last publications before he died was the Introduction to the seven-volume Dictionnaire géographique et administratif edited by Paul Joanne. Reclus's introductory volume (1905) was actually the last volume published in the series, which began in 1890. It is a revised version of the earlier geographical introductions that Elisée and Elie had done for Joanne père in the 1860s. Although Paul's name does not appear in the 1905 volume, he was actually responsible for more than 70 percent of the book.[41]

Another project which was finished by Paul but which bears only his uncle's name was Les Volcans de la terre, which was published in three parts from 1906 to 1910. Elisée's early interest in volcanoes had been underscored by his ascent of Etna in 1865, at which time he stated reasons for studying volcanic forces and other natural phenomena:

> If we can study these forces and understand them, perhaps we can neutralize or harness them. In order to triumph over nature, it is first necessary to understand her.[42]

In La Terre Reclus said that it might be possible to discover "the rhythm of the great volcanic revolutions." It is of interest to note that

[121]

the French traveler Gabriel Bonvalot gave the name Volcan Reclus to a peak which he had "discovered" in Tibet in 1889.[43]

Reclus's interest in volcanoes was revived in 1902 by the eruption of Mont Pelée in Martinique. In 1903 he proposed an "Atlas of Volcanoes," which would include a world map at the scale of 1:40,000,000 and ten regional maps at 1:2,000,000. It would be a large-sized volume of about three hundred pages and would be published by the Société belge d'astronomie. It was actually published in three fascicules totalling 515 pages. Only Southwest Asia and Southern Europe were covered, with heavy emphasis on Italy. The work was praised for the "neatness, sobriety, and elegance" of the maps and for its useful factual information, even though it was generally devoid of critical or scientific analysis.[44]

Although Reclus had suffered from angina from 1880 onward, he maintained a vigorous routine almost to the end. He declined noticeably in the last two years of his life. His health seemed to be poorer than Elie's in the six months before the latter's death, and nephew Paul claimed that he was suffering more from arteriosclerosis than from angina. Elisée's last public speech was in Paris in February 1905, after the beginning of the abortive revolution in Russia. He rose to speak in support of the revolution, but he had to sit down after only five minutes because of his heart.[45]

During the last few weeks of his life, Elisée knew that death was approaching, and so he tried to tie up all the loose ends. His sister Louise was at hand, and brother Paul, the surgeon, came up from Paris to be with Elisée at the end, but he left strict instructions that only nephew Paul was to accompany his body to the cemetery. On the last day of his life Reclus received word of the revolt of the sailors on the Russian ship "Potemkin," and his last words were reported to have been, "The Revolution! At last, the Revolution has come!" If this is true, he died happily, because this event would have been for him the seventh fanfare of the trumpets of Jericho. He expired at Mme. de Brouckère's country place at Thourout on the night of 4-5 July 1905 and then was interred with Elie in the cemetery in Ixelles (Brussels). Their modest headstone contrasts with the rather more elaborate stone adorning the adjacent grave of a baby girl, the daughter of a dentist from Pasadena, California![46]

Chapter Seven

Legacy

Without disputing Elie Faure, it is obvious that the beacon has had greater influence or radiance than the star. A person's legacy can be seen in the institutions that he builds, in his writings or other creative works, and in his biological and intellectual progeny. In the case of Elisée Reclus, the institution that he created, the Institut géographique de Bruxelles, was ably managed by his nephew until it was silenced by the "guns of August." The Institut did not produce graduates of the usual sort, and so Reclus was not able to establish a real school as Vidal was fortunate in doing in Paris. However, he left behind an enormous and coherent body of literature, perhaps the largest corpus of materials written by any geographer of the modern era, and, although his once-widespread fame has naturally diminished with the passage of time, it may be safe to say that he still has an edge on the "mainstream" geographers of his day, such as Friedrich Ratzel, Ferdinand von Richthofen, and perhaps even Paul Vidal de la Blache himself. It is also possible that there is greater likelihood of the revival of interest in Reclus's writings in the future, partly because of his radicalism and partly because he produced works of universal appeal. If this modest biography contributes to this renascence, then its author will feel that his efforts have been worthwhile.

Reclus was fortunate in that he was a member of a large and devoted family which has always worked to keep his name alive, in France and in the world generally. Of his nine younger siblings, who died between 1910 (Löis) and 1937 (Ioanna), probably Onésime (died 1916) and Louise (died 1917) were the most significant in aiding

[123]

Elisée. A prolific geographer himself, Onésime was of great help to Elisée as his agent in Paris for many years. Louise was her brother's secretary and general aide in the Brussels years. She is responsible for preserving Elisée's correspondence and for publishing large amounts of it. After her death the manuscript materials which she collected were deposited in the Bibliothèque Nationale, where they are kept in eleven folio volumes.[1]

Elisée had wisely chosen his nephew Paul to carry on his work. Paul worked very diligently in directing the Institut géographique and in completing Elisée's unfinished works. Paul always took great pains to conceal his own contributions, but he deserves a large measure of credit for the enlargement of Elisée Reclus's reputation. Paul did not return to academic life after World War I but instead moved into a house that was left to him by his aunt Noémi Mangé in Domme on the Dordogne River about a hundred kilometers east of Sainte-Foy. There he remained, except for numerous forays to Paris and especially to Montpellier, where he managed the Collège des Ecossais after the death of Patrick Geddes in 1932. Paul remodeled an old windmill in Domme to serve as a regional museum much like Geddes' Outlook Tower in Edinburgh. A few items are still on display in Domme in a small Musée Paul Reclus, but the books were moved to the Geographical Institute of the University of Tours in the late 1960s. Much material still resides in Paris with Paul's son Jacques, a distinguished Sinologist.

While living in Domme in reduced circumstances in the 1920s, Paul nevertheless managed to provide unstinted assistance to Max Nettlau and Joseph Ishill on their Reclus biographies. Ishill (1888-1966) was a Romanian-born American libertarian who printed books of unusual beauty on a small handpress at his home in New Jersey. After Ishill sent Paul Reclus a copy of a book of essays on Kropotkin that he had published in 1923, Paul began in earnest to aid Ishill and Nettlau in accumulating materials for the book on Elie and Elisée Reclus which Ishill printed in 1927. Paul not only wrote the longest and most important essay in the book, "A Few Recollections on the Brothers Elie and Elisée Reclus," but he assisted the enterprise in numerous other ways. The greater length and value of Ishill's Reclus book, as compared with his previous book on Kropotkin, is largely attributable to the efforts of Paul Reclus. In 1924 Nettlau started to write a small piece on Elisée Reclus, whom he had known in the period 1891-1903, but he soon extended it to book length and published it in 1928. He had put off writing such a book previously because it was always assumed that Jacques Mesnil would write the definitive biography of Elisée Reclus. The 1929 Spanish edition of Nettlau's biography is superior to the original German edition, because it incorporates materials which Paul Reclus had provided too late for the initial publication. Paul was also then working on a three-volume abridgement of Elisée's *L'Homme et la terre*, which was published in 1930-1931. This represented virtually his only major contribution to academic or professional geography after he left Brussels in 1914, although he always considered himself to be a geographer. As he said in an autobiographical piece in the anarchist newspaper *Plus Loin* in 1938, the basic program of geography—to know the Earth and to understand how Man can fashion it for his use and organize himself in order to inhabit it intelligently—involves a synthesis, not only of the kinds of geography that one is traditionally taught, but also of engineering studies, the regional museum, and the anarchist mentality.

During the 1930s Paul expanded the article that he had written for Ishill's book, but, unfortunately, he did not publish his Reclus biographical materials before he died in 1941. They were finally gathered together by his sons and their associates and published in 1964 under the revealing title, *Les Frères Elie et Elisée Reclus: ou, du protestantisme à l'anarchisme*.[2]

Elisée Reclus has left numerous descendants through his daughters, the younger of whom, Jeannie Cuisinier (later Ostroga), died in Menton in 1897, but the older, Magali Régnier, lived on until 1951. Although not all members of this remarkable *tribu* are still found on the left side of the political spectrum, they nevertheless demonstrate unusual family solidarity and are all devoted to the spirit of their illustrious ancestor. Elisée was closer to the Cuisiniers because they lived near him after the death of Léon Cuisinier, Jeannie's husband, in 1887, but he tried to visit the Régniers in Algeria as often as possible. Elisée had perhaps a special fondness for the oldest grandchild, Louis Cuisinier (1883-1952), who displayed keen interest in travel and even provided some photographs for *L'Homme*. Louis reminded Elisée of his own youth fifty years earlier. One of Elisée's great-granddaughters, Magali Cuisinier's daughter, Jeannie Collin, married Patrick Geddes's son Arthur, a geographer at the University of Edinburgh, and so the strong bond between the Reclus and Geddes families was made permanent. In 1970 one of the members of the next generation, Anne Geddes Shalit, made news by successfully challenging the requirements for Israeli citizenship. The Reclus spirit remains undiminished![3]

As one of that mighty band of French socialist writers of the nineteenth century, Elisée Reclus has had a general, though immeasurable influence on the political activists of the modern era. He might well have been mentioned by Edmund Wilson in that host of writers, mostly French, who propelled Lenin toward the Finland Station in 1917. I do not know whether Lenin actually read any of Reclus's works, which had achieved wide popularity in Czarist Russia, but it is said that Stalin had derived his basic notions about environmental influences from Reclus. The young Chinese intellectuals who were prominent in Paris after 1905 were avid readers of Reclus and Kropotkin, and Mao Tse-tung himself was profoundly influenced by anarchism, although he was frustrated in his hope of going to Paris. Reclus's fame as an anarchist has been perhaps even more enduring in Latin America than elsewhere, and his name has been specifically invoked in the political education of the Mexican revolutionaries, the Flores Magon brothers, who in turn influenced Emiliano Zapata. The famous Spanish writer Vicente Blasco Ibáñez was instrumental in the publication and wide Iberian and Latin American dissemination of several of Reclus's works, including a six-volume *Novísima geografía universal* (Valencia and Madrid, 1906-1907).[4]

Although there are few people now alive who actually knew Elisée Reclus—and not many more who can claim to have read a large portion of his works—there seems to be a renewed interest in his life and writings, just as there is in his friends Patrick Geddes, Peter Kropotkin, and Ferdinand Domela Nieuwenhuis.[5] In the English-speaking world there is greater interest in Kropotkin and Geddes at the moment, undoubtedly because of the greater accessibility of their books, but Reclus is often mentioned in the same breath. The aged American philosopher Will Durant still remembers the excitement of his encounter with the works of Elisée Reclus and Peter Kropotkin at the Ferrer Center in New York in 1912 and especially of his meeting with Kropotkin in London in the same year. Kropotkin may be better known than Reclus in Great Britain and North America simply because his title of Prince seems to stick in everyone's mind. In a short story by the Italian writer Italo Calvino, there is a passage, "In the evening my father reads aloud some books of Elisée Reclus," but the published English translation reads, "In the evening, my father reads out loud from Kropotkin." The translators must have been aware that the works of Reclus and Kropotkin were similar but that the latter name would be more familiar to readers of English.[6]

Elisée Reclus is often categorized as a "literary," rather than as a "scientific," geographer—a characterization intended to praise the quality of his writing but noting his relative deficiency in scientific rigor. A British geographer, R. J. Harrison Church, has called Reclus "the most famous representative of the best kind of descriptive geography." "He did a very great deal," said Harrison Church, "to induce a new interest in geography in France—albeit of a rather literary kind." André Meynier has echoed the usual praise of Reclus's literary skill, but he feels that Reclus always went beyond mere description and attempted to move to a higher level of explanation:

> His charm is owed essentially to his literary qualities. His descriptions have a force of evocation that has not aged. But if Reclus was the best representative of descriptive geography, one would be wrong in seeing in him only his literary talent. Each time that he could, he tried to explain what he had seen.[7]

The doyen of French human geographers, Aimé Perpillou, who worked with Paul Reclus on the revision of Elisée Reclus's L'Homme et la terre fifty years ago, has succinctly characterized the virtues and the enduring quality of Reclus's work:

> While Elisée Reclus's methods and geographical concepts have aged, he is still read today as a witness of an era and of a generation of geographers of whom he was one of the greatest. His descriptions are vivid and exact; his theories remain in great part valid. . . . His work is, furthermore, in France, the last geographical encyclopedia, the last "Géographie universelle" written by a single author, with, consequently, a unity of thought and method, which make it a vigorous synthesis of the geographical knowledge of a century ago.[8]

Although Reclus's method and style may be démodé, his works are still cited for descriptions of places as they appeared a century ago. Indeed, Charles Fisher has indicated that the *NGU* might be increasingly valuable to historians and historical geographers for its wealth of detail for the late nineteenth century, "which is in danger of becoming the forgotten period between historical and contemporary geography." Even Reclus's contemporaries criticized his works for their deficiency in analytical rigor, but all praised his literary style and the warmth of his sentiments. In an obituary the Italian geographer Olinto Marinelli wrote an accurate assessment of Reclus's geography that remains valid today:

> The colossal work [*NGU*] cannot be said to be new, either for singularity of plan, or for grandiosity of construction, or for particular originality of vision . . .
>
> One can disagree with him; his ideas can be called utopian; but no one can escape from feeling a keen attraction for his writings and a great sympathy for their author. . . .
>
> Neither the optimistic nor the artistic tendencies do damage to the scientific value of his books. . . .
>
> In profundity of thought, in acuteness of criticism, and in originality of ideas, Elisée Reclus was surpassed by some of his contemporaries. A few months have passed since the greatest of these died: Friedrich Ratzel. . . . No matter how grand it is, Reclus's work remains closed within itself; it can be imitated but not perfected; it can stimulate admirers of geography but not form new geographers. He did not leave true students. He did not indicate new roads to travel. Art is neither taught nor learned; and his work is more a work of art than of science.[9]

Even though Reclus's art cannot be taught or passed on directly to students, his life can be an inspiration to them. In 1916 another Italian geographer, Luigi Filippo de Magistris, said that Elisée Reclus will be appreciated more and more as time goes on, because people will not admire the scientist who shuts himself up in his laboratory so much as the one who brings the world into his work. The popular "discovery" of ecology and the new social geography in recent years has given a special cachet to radical or activist geographers. In raising the call for a more "relevant" geography, they are not entirely innovative, because the anarchist geographers were saying essentially the same things a century ago. This reminder is not meant to deprecate the earnestness of the present-day geographers who want to be social activists but merely to ask them to explore their intellectual heritage.[10]

André Meynier has asked why Elisée Reclus is not better known today, and he suggests that three factors might be responsible for Reclus's fading reputation: his "Marxist" (*sic*) ideas, his long years of exile, and his easygoing but undemanding writing style. Reclus was not, of course, a Marxist, although he shared many of Marx's basic ideas. In any event, his advanced political opinions did not appear to harm his reputation with his largely bourgeois readership, in his own or succeeding generations. Reclus's years of true exile, 1872-1879, were highly productive, although Meynier was correct in saying that he could not have had the same influence as a Parisian professor while operating from a base in Switzerland or later from Brussels. Reclus's writing, said Meynier, had the tone of a pleasant conversation and lacked the didactic or structured style that the French, being carte-sians, seem to prefer. Meynier credits Reclus with creating a favorable atmosphere for geography but not with a commanding rôle in directing the course of its evolution.[11]

Béatrice Giblin has said that Elisée Reclus has been "erased" from the history of geography in France because people were afraid of his anarchism and because he was not in the mainstream of the new university-centered geography. The latter explanation would be more admissable than the former, but, even more basically, I do not accept the initial premise that Reclus has been erased. Obviously, his fame diminished after his death, but that was due simply to the passage of time and not to conspiracy. How many of Reclus's contemporaries are now better known? French geographers have not forgotten Reclus. They seem to have placed his life in proper perspective, neither unduly glorifying nor denigrating him. There is no reason to suspect that Reclus's geography has been suppressed out of fear. I agree with Giblin's view that Elisée Reclus's geography was a "dangerous" science because he showed that the earth can support everyone and because he pointed out past errors in the management of people and resources. But geographers—and all other scholars—must always be "dangerous" or "subversive," in the sense that Paul Sears has called ecology "the subversive science," because it is precisely their task to examine all policies critically, with an eye toward their improvement. Scholars are expected to expand the world-view of their countrymen, who have less time for intellectual activities. To do otherwise would be unpatriotic or truly subversive, in that we would be subverting our educational mission.[12]

Outside France, Reclus's geography has fared better in Britain than elsewhere, partly because the French style of geography has traditionally found a warm reception there and partly because Patrick Geddes was successful in introducing Reclus's works to British geographers. Geddes also recommended Reclus's writings to his American disciple Lewis Mumford, but, although Mumford's interests are quite close to those of Reclus, he has not cited him extensively. Furthermore, Mumford does not have close ties with American geographers, although he is much admired by them. American links, in the past at least, have been more with Germany than with France or even Great Britain. But now, on both sides of the Atlantic, Reclus's geography may find greater favor as geographers are groping toward a more "humane" geography. They could seek no better mentor than Elisée Reclus.

Descriptions of Elisée Reclus in 1871 and 1890

Physical descriptions of Elisée Reclus were given in the *Jugement* of 15 November 1871 announcing his sentence of deportation and also on the passport that he obtained in Oran, Algeria on 6 March 1890. The former can be found in the Reclus Papers in the Bibliothèque Nationale (NAF 22911: 82), and the latter is in the possession of Mme. Louise Rapacka.

1. *Jugement* of the 7th Council of War against "RECLUS, Elysée, Jacques, écrivain géographe"

Height	one meter 600 millimeters [5′3″]
Hair and Eyebrows	*chestnut-brown*
Forehead	*high*
Eyes	*blue*
Nose	*average*
Mouth	*average*
Chin	*rounded*
Face	*long*
Complexion	*ordinary*

2. Passport (6 March 1890)

Age	*60 years* [sic—actually Reclus would not turn 60 until 15 March 1890]

Height	one meter 65 centimeters [5'5''—probably more nearly accurate than the figure in 1871]
Hair	*graying*
Forehead	*ordinary*
Eyebrows	*chestnut-brown*
Eyes	*blue*
Nose	*average*
Mouth	*average*
Beard	*graying*
Chin	*rounded*
Face	*oval*
Complexion	*ordinary*

In the notes for his projected biography of Elisée Reclus, George Engerrand also recorded his height as one meter 65 centimeters. Engerrand said that Reclus often regretted his short stature, because, he said, there are times when it is necessary "to shove a coward up against the wall."

Notes

Notes to Preface

1. J. K. Wright, *Human Nature in Geography* (Cambridge, Massachusetts: Harvard University Press, 1966), 10.

2. Kenneth Rexroth, "Revolution Now!," *San Francisco*, vol. 11, no. 12 (December 1969), 20; cf. Carl Landauer, *European Socialism* . . ., volume 1 (Berkeley and Los Angeles: University of California Press, 1959), ix.

3. Terry N. Clark, *Prophets and Patrons: The French University and the Emergence of the Social Sciences* (Cambridge: Harvard University Press, 1973); Benjamin Harrison, "Gabriel Monod and the Professionalization of History in France, 1844-1912," Unpublished Ph.D. thesis in History, University of Wisconsin, 1972; William R. Keylor, *Academy and Community: The Foundation of the French Historical Profession* (Cambridge: Harvard University Press, 1975); and Vincent R. H. Berdoulay; "The Emergence of the French School of Geography (1870-1914)," Unpublished Ph.D. thesis in Geography, University of California, Berkeley, 1974.

4. Franz Schrader, "Elisée Reclus," *La Géographie*, vol. 12, no. 2 (15 August 1905), 85.

5. [Georges de Nouvion] "Etrennes 1887," *Revue bleue* (*Revue politique et littéraire*), 3rd series, 23rd year, 2nd semester, no. 26 (25 December 1886), 813.

Notes to Chapter One

1. Letter from Bigelow to Huntington, 17 June 1868, in John Bigelow, *Retrospections of an Active Life,* vol. 4 (Garden City, N.Y.: Doubleday, Page & Co., 1913), 185-186.

2. Paul Reclus, "Biographie d'Elisée Reclus," *Les Frères Elie et Elisée Reclus; ou, du protestantisme à l'anarchisme* (Paris: les Amis d'Elisée Reclus, 1964), 15. Hereafter cited as P. Reclus, *Frères*. The "Biographie" was written by Paul Reclus (1858-1941) in 1939.

3. The Reformed Church was suppressed between 1685 (Revocation of the Edict of Nantes) and 1787, when an edict of toleration was issued by Louis XVI. Services had to be held surreptitiously in caves, or in the woods, or in other such places—literally in the desert. The Church was thus called the "Church of the Desert," and the ministers were the "pastors of the desert." See Pierre Larousse, *Grand dictionnaire universel du XIXe siècle*, vol. 6 (Paris: Administration du Grand dictionnaire universel, n. d.), 544-545.

4. Jean Corriger, "Notice historique sur Sainte-Foy-la-Grande," in P. Reclus, *Frères*, 204.

5. There are interesting parallels between the *tribu* Reclus and the *tribu* Monod, the family of the great historian Gabriel Monod. See Benjamin Harrison, "Gabriel Monod and the Professionalization of History in France, 1844-1912," unpublished Ph.D. dissertation, University of Wisconsin, 1972. Just as Jacques Reclus had eleven children, many of whom made a significant mark, Gabriel Monod's grandfather had twelve offspring, who were to become known as "les douze" and develop what was recognized as a true *tribu* by the end of the nineteenth century. The Protestant origins of both families are noteworthy.

6. P. Reclus, *Frères*, 16-17; Daniel Robert, *Les Eglises Reformées en France (1800-1830)* (Paris: Presses Universitaires de France, 1961), 405-406, 484, 570, 581-582; Daniel Robert, *Textes et documents relatifs à l'histoire des églises réformées en France (période 1800-1830)* (Geneva and Paris: Droz and Minard, 1962), 353-357; Daniel Robert, "Note sur l' 'Affaire Henriquet-Reclus' et les origines de l'église 'libre' de Sainte-Foy," Fédération historique du Sud-Ouest, *Sainte-Foy-la-Grande et ses alentours* (Bordeaux: Editions Bière, 1968), 113-118; Agnès de Neufville, *Le Mouvement social protestant en France depuis 1880* (Paris: Les Presses Universitaires de France, 1927), 38.

7. For biographical information on Jacques Reclus, see [Elisée Reclus] *Elie Reclus, 1827-1904* (Paris: L'Emancipatrice, n. d. [1905]), 7-9; R. Marzac, "Les Reclus," *Figaro* (30 July 1894), 1; Robert de Bonnières, *Mémoires d'aujourd'hui: Troisième série* (Paris: Paul Ollendorff, Editeur, 1888), 199; Paul Reclus, "A Few Recollections on the Brothers Elie and Elisée Reclus," *Elisée and Elie Reclus: In Memoriam*, ed. by Joseph Ishill (Berkeley Heights, New Jersey: The Oriole Press, 1927) (hereafter cited as Ishill, *Reclus*), 7-8; Peter Kropotkin, "Elisée Reclus," *Les Temps Nouveaux*, 11th year, no. 11 (15 July 1905), 1; and N. Weiss, "Les Protestants en France . . .," *Bulletin de la Société d'histoire du protestantisme français*, vol. 61, no. 1 (January-February 1912), 11-12. Many clues to Jacques Reclus' personality are found in the little book that he published anonymously in 1858: *Scènes d'une pauvre vie* (Pau: Typographie et lithographie Veronese). A touching obituary notice was published in the *Mercure d'Orthez* (12 April 1882).

8. P. Reclus, *Frères*, 11, 17-18; [Elisée Reclus] *Elie Reclus, op. cit.*, 8-9; Kropotkin, *op. cit.*; Marzac, *op. cit.*; Max Nettlau, *Elisée Reclus, Anarchist und Gelehrter (1830-1905)* (Berlin: Verlag "Der Syndikalist," Fritz Kater, 1928) (hereafter cited as Nettlau, *Elisée Reclus*), 12, 102; and Gabriel Astruc, "La Famille Reclus," *L'Illustration*, vol. 102, no. 2652 (23 December 1893), 569.

9. Nettlau, *Elisée Reclus*, 15-16; official transcripts of birth records in Sainte-Foy and Orthez made for George Engerrand in 1961.

10. Corriger in P. Reclus, *Frères*, 204; Maurice Beresford, *New Towns of the Middle Ages: Town Plantation in England, Wales and Gascony* (London: Lutterworth Press, 1967), 146 *et passim*; Adolphe Joanne, *Itinéraire général de la France: La Loire et le Centre* (Paris: Hachette, 1868), 482; 1780 population figure from Municipal Archives of

Sainte-Foy transmitted by the Mayor to Mme. Jean Corriger in letter of 5 December 1975.

11. Elisée Reclus's birth certificate is recorded in the Hôtel de Ville of Sainte-Foy-la-Grande in a volume labeled "Naissances. 1833 [sic] à 1842," in a section labeled "An 1830. Registre des naissances. Commune de Ste. Foy la Grande." I examined the birth register in July 1967 in the company of the late Arthur Geddes. I noted that a child born in Sainte-Foy at nearly the same time as Elisée Reclus was identified as the son of an "Irish gentleman." Jokingly, I asked Geddes if there was not an inherent paradox in that phrase, if it were possible to be both Irish and a gentleman, and he drew himself up and said that it would be impossible *not* to be both.

12. Franz Schrader, "Géographie," *Dictionnaire de pédagogie et d'instruction primaire*, ed. by F. Buisson, part 1, vol. 1 (Paris: Hachette, 1882), 1155.

13. The Neuwied years of the Reclus brothers and George Meredith are described in [Elisée Reclus] *Elie Reclus, op. cit.*, 11-14; P. Reclus, *Frères*, 18-19; Nettlau, *Elisée Reclus*, 20-22; René Galland, *George Meredith: Les cinquantes premières années (1828-1878)* (Paris: Les Presses Françaises, 1923), 22-38; J. A. Hammerton, *George Meredith: His Life and Art in Anecdote and Criticism* (Edinburgh: John Grant, 1911), 4-5; Mona Mackey, *Meredith et la France* (Paris: Boivin et Cie., Editeurs, 1937), 11-12; and William Meredith, ed., *Letters of George Meredith*, vol. 1 (London: Constable and Company, Ltd., 1912), 3.

14. Elisée's baccalaureate diploma is preserved with his papers in the Manuscripts Department of the Bibliothèque Nationale (N.A.F. 22909: feuillet 7). Henceforth the Reclus papers in the Bibliothèque Nationale (N.A.F. 22909-22919) will be cited simply as BN 22909, etc.

15. The details of the Reclus brothers' lives in Sainte-Foy and Montauban in the 1840s can be found in [Elisée Reclus] *Elie Reclus, op. cit.*, 15-20; Ishill, *Reclus*, 10-11, 24-25; Nettlau, *Elisée Reclus*, 23-28; and P. Reclus, *Frères*, 19-21.

16. Elisée Reclus's period at the University of Berlin is described in his letters to his parents. See Elisée Reclus, *Correspondance*, vol. 1 (Paris: Schleicher Frères, 1911), 31-39 (the 3-vol. *Correspondance* will be hereafter cited as *Corr.*); Nettlau, *Elisée Reclus*, 35; P. Reclus, *Frères*, 21; BN 22909: 8, 22910: 15-18. Reclus's matriculation certificate ("Elisaeus Reclus, Gallus, theologiae studiosus") is dated 1 February 1851. It is preserved with his papers in the Bibliothèque Nationale (BN 22909: 8). I am indebted to Dr. Gerhard Engelmann for information about Reclus's final period in Berlin.

It is interesting, but perhaps not significant, that Karl Marx had taken Ritter's course in 1838. See Auguste Cornu, *Karl Marx et Friedrich Engels*, vol. 1 (Paris: Presses universitaires de France, 1955), 133.

17. Patrick Geddes, "The Education of Two Boys," *The Survey*, vol. 54, no. 11 (1 September 1925), 574-575.

18. The essay was printed in *Le Libertaire* (Paris) in 1925. See Clara Mesnil, "Souvenirs sur Elisée Reclus," *Maintenant*, no. 2 (1946), 254-255.

19. The events in Orthez in December 1851 are briefly described in [Elisée Reclus] *Elie Reclus, op. cit.*, 23; Nettlau, *Elisée Reclus*, 49-50; and P. Reclus, *Frères*, 23. In 1977, at my request, Mme. Michelle Imbert of Pau searched the Orthez newspapers and archives in vain for any references to the Reclus brothers in connection with the town hall incident, which is perhaps not surprising, because such information was suppressed by the officials.

Notes to Chapter Two

1. Although there is no exact French equivalent of the German term "Wander-jahre," perhaps the sense could be communicated by the familiar expression, "Les voyages forment la jeunesse," or by a new construction such as "errance estudiantine." I should like to thank Mme. Gemma Raso for this suggestion.

2. Reclus's life in London is treated in *Corr.*, I, 44-57; P. Reclus, *Frères*, 24-25; Nettlau, *Elisée Reclus*, 50-52; etc.

3. "Dame Rebecca West Talks to Anthony Curtis about Social Improvements and Literary Disasters," *The Listener*, vol. 89, no. 2290 (15 February 1973), 211; Letters from Rebecca West, 2 May and 3 September 1973. For mention of the Fairfield brothers by Elisée Reclus, see especially *Elie Reclus*, 25.

4. *Corr.*, I, 479, II, 221n, 251-255; Ishill, *Reclus*, 93-100; Nettlau, *Elisée Reclus*, 52.

5. Ralph Hyde, "Mr. Wyld's Monster Globe," *History Today*, vol. 20, no. 2 (February 1970), 118-123; "Mr. Wyld's Large Model of the Earth," *The Illustrated London News*, vol. 18, no. 475 (22 March 1851), 234; "Mr. Wyld's Model of the Earth," *Ibid.*, vol. 18, no. 491 (7 June 1851), 512; Elisée Reclus, "A Great Globe," *Geographical Journal*, vol. 12, no. 4 (October 1898), 403; G. S. Dunbar, "Elisée Reclus and the Great Globe," *Scottish Geographical Magazine*, vol. 90, no. 1 (April 1974), 57-58.

6. Elisée in Ireland is treated in *Corr.*, I, 58-60; P. Reclus, *Frères*, 26; and Nettlau, *Elisée Reclus*, 55-56.

7. Reclus's life in Louisiana is described in *Corr.*, I, 68-111; P. Reclus, *Frères*, 27-30; and Nettlau, *Elisée Reclus*, 58-69.

8. Information on the Fortier and Aime families can be found in Estelle M. Fortier Cochran, *The Fortier Family and Allied Families* (N. p., n. d. [San Antonio, Texas; The Author, 1963]; Roulhac B. Toledano, "Louisiana's Golden Age: Valcour Aime in St. James Parish," *Louisiana History*, vol. 10, no. 3 (Summer 1969), 211-224; and R.P. Martin, Jr., "The Plantation Mansion and Estate of Valcour Aime, St. James Parish; or, A Brief Discussion of the Historical and Architectural Aspects of the Plantation Known as the St. James Refinery," unpublished typescript (May 1968) in Special Collections Division, Howard-Tilton Memorial Library, Tulane University.

For descriptions of the agricultural enterprises of Valcour Aime and the Fortier brothers in the 1850s, see *Plantation Diary of the Late Mr. Valcour Aime . . .* (New Orleans: Clark & Hofeline, 1878); P.A. Champomier, *Statement of the Sugar Crop Made in Louisiana, in 1852-53 . . .* (New Orleans: Cook, Young & Co., 1853); and T.B. Thorpe, "Sugar and the Sugar Region of Louisiana," *Harper's New Monthly Magazine*, vol. 7, no. 42 (November 1853), 758-759.

Reclus's letter to Clara Mesnil (25 October 1904) was published in part in *Corr.*, III, 291-293, but the section concerning Anna Fortier was left out, perhaps intentionally suppressed by the editor, Paul Reclus.

9. Lafaye is mentioned in *Corr.*, I, 70, 79, 102. See also "Registre du Comité Médical de la N^{ell} Orléans," Unpublished manuscript in Rudolph Matas Medical Library, School of Medicine, Tulane University, p. 180; U.S. Census, 1850, Louisiana, St. James Parish, Film 208 in Louisiana Collection in Special Collections Division, Howard-Tilton Memorial Library, Tulane University; and Cecile Willink, ed., "An Old Lady's Gossip of Life in Louisiana in the Middle of the Last Century," *Louisiana Historical Quarterly*, vol. 6, no. 3 (July 1923), 385-386.

10. Elisée Reclus, "Le Mississipi, Etudes et souvenirs.—I.—Le Cours supérieur du fleuve," *Revue des Deux Mondes*, vol. 22 (15 July 1859), 274-275.

11. Elisée Reclus, "De l'esclavage aux Etats-Unis.—I.—Le code noir et les esclaves," *Ibid.*, vol. 30 (15 December 1860), 892.

12. *Ibid.*, vol. 22 (1 August 1859), 625.

13. *Corr.*, I, 96.

14. *Ibid.*, I, 106.

15. *Ibid.*, I, 109.

16. *Ibid.*, I, 110.

17. Elisée Reclus, "Lettres d'un voyageur," *L'Union*, vol. 1, no. 7 (7 February 1857), [4]; *Ship Registers and Enrollments of New Orleans, Louisiana*, vol. 5 (1851-1860) (Work Projects Administration, Survey of Federal Archives in Louisiana) (University, La.: Hill Memorial Library, Louisiana State University, 1942), 209.

18. *L'Union*, vol. 1, no. 16 (16 February 1857), [4]; Paul Harvey and J. E. Heseltine, eds., *The Oxford Companion to French Literature* (Oxford: Clarendon Press, 1959), 370.

19. Elisée's adventures in Colombia are chronicled in *Corr.*, I, 112-167, and especially in his book, *Voyage à la Sierra-Nevada de Sainte-Marthe: Paysages de la nature tropicale* (Paris: Librairie de L. Hachette et Cie., 1861), first published serially in *Revue des Deux Mondes*, vol. 24 (1 December 1859), 624-661; vol. 25 (1 February 1860), 609-635; vol. 26 (15 March 1860), 419-452; and vol. 27 (1 May 1860), 50-83. Cf. Gabriel Giraldo Jaramillo, *Bibliografía colombiana de viajes* (Biblioteca bibliografía colombiana, II) (Bogotá: Editorial ABC, 1957), 183-184. *Voyage* was Reclus's second book. It is a descriptive account of northeastern Colombia and his experiences there. It would appear that some names were altered and some events might have been romanticized or slightly fictionalized, but on the whole it seems to be an authoritative work.

20. *Revue des Deux Mondes*, vol. 27 (1 May 1860), 80; BN 22917: 67-68; Thomas D. Cabot, "The Cabot Expedition to the Sierra Nevada de Santa Marta, Colombia," *Geographical Review*, vol. 29, no. 4 (October 1939), 609-611; James R. Krogzemis, "A Historical Geography of the Santa Marta Area, Colombia," Unpublished Ph.D. dissertation, University of California, Berkeley, 1967.

Notes to Chapter Three

1. BN 22909: 9, 22-23; *Corr.*, I, 183-184; P. Reclus, *Frères*, 43-44; Ishill, *Reclus*, 11-12.

2. BN 22910: 88-90; *Corr.*, I, 169-172.

3. *Ibid.*

4. *Corr.*, I, 173-174, 181-182; *Bulletin de la Société de géographie*, series 4, vol. 16 (July-August 1858), 114, 119.

5. Georges Pariset, "La 'Revue Germanique' de Dollfus et Nefftzer . . .," *Revue germanique*, vol. 2, no. 1 (January 1906), 33.

6. *Revue germanique*, vol. 8, no. 11 (1859), 241-267; reprinted with Ritter's "Introduction à la géographie générale comparée," with commentary by Georges

Nicolas-Obadia (*Cahiers de géographie de Besançon*, 22) (Paris: Les Belles Lettres, 1974).

7. L. Reynaud, *L'Influence allemande en France au XVIIIe et au XIXe siècle* (Paris: Librairie Hachette, 1922), 150.

8. Jean Mistler, *La Librairie Hachette de 1826 à nos jours* (Paris: Hachette, 1964), 257-262.

9. Nettlau, *Elisée Reclus*, 88n, 89; Mistler, *op. cit.*, 261; Franz Schrader, "Emile Templier," separately paged obituary bound in at the end of vol. 61 (1st semester 1891) of *Le Tour du monde*; article on Adolphe Joanne in *Grand dictionnaire universel du XIXe siècle*, vol. 9 (Paris, 1873), 991. Joanne's praise of Reclus was published in *Itinéraire général de la France*, vol. 3, *Les Pyrénées et le réseau des chemins de fer du Midi et des Pyrénées* (Paris: Hachette, 1862), xv.

10. Adolphe Joanne, *Itinéraire descriptif et historique de la Suisse, du Jura français, du Mont-Blanc et du Mont-Rose*, 3rd ed. (Paris: Hachette, n. d. [1859]), xiii-xiv. For a bibliography of the enormous literary output of the Joannes and their allies, see *Catalogue général des livres imprimés de la Bibliothèque Nationale*, vol. 78 (Paris, 1923), cols. 118-294.

11. Elisée Reclus, "Le Littoral de la France.—II.—Les Landes du Médoc et les dunes de la côte," *Revue des Deux Mondes*, vol. 46 (1 August 1863), 702.

12. Elisée Reclus, "Du Sentiment de la nature dans les sociétés modernes," *Revue des Deux Mondes*, vol. 63 (15 May 1866), 363-365, 379.

13. Elisée Reclus, "L'Homme et la nature.—De l'action humaine sur la géographie physique," *Revue des Deux Mondes*, vol. 54 (15 December 1864), 763.

14. *Ibid.*, 770-771.

15. Letters from John Bigelow to William Seward, 29 May and 3 July 1863, Microcopy No. T-1, Despatches from United States Consuls in Paris, 1790-1906, Roll 13, vol. 13, U.S. National Archives, Record Group 84; John Bigelow, *Retrospections of an Active Life*, 5 vols. (New York, 1909, and Garden City, 1913), II, 25-26, 51, 88-89, V, 26-28. Although a number of sources say that Elisée Reclus refused an offer of a large sum of money by the American minister in Paris for writing pro-Union articles during the American Civil War, I am quite certain that they were actually referring to Bigelow's personal gift in 1872, which was accepted by Reclus but not for himself. See Jean Steens, "Profils socialistes," *Le Correspondant*, n. s. vol. 175, no. 6 (25 June 1903), 1176, and Maurice Peyrot, "Elisée Reclus," *La Nouvelle Revue*, vol. 50, no. 1 (1 January 1888), 172.

16. Bigelow, *Retrospections*, II, 527; *Corr.*, I, 244-246; Elisée Reclus, "La Sicile et l'éruption de l'Etna en 1865," *Le Tour du monde*, vol. 13 (1866), 358.

17. Reclus, "La Sicile . . .," *Ibid.*, 353-416 (quotation on pp. 387-388); Reclus, *Les Volcans de la terre*, 3 parts (Brussels: Société belge d'astronomie, de météorologie et du physique du globe, 1906-1910).

18. Ishill, *Reclus*, 197-208; George Woodcock, *Anarchism: A History of Libertarian Ideas and Movements* (Cleveland and New York: Meridian Books, The World Publishing Company, 1962), 183.

19. Archives Librairie Hachette, Contract between Hachette and Reclus dated 24 December 1862.

20. Elisée Reclus, *La Terre*, vol. 2 (Paris: Hachette, 1869), 728-729.

21. *La Philosophie positive*, vol. 2, no. 4 (January-February 1868), 160; *Annales des voyages . . .*, vol. 197 (February 1868), 241; *Ibid.*, vol. 200 (December 1868), 307-308; *Athenaeum*, no. 2109 (28 March 1868), 463-464.

22. [Oscar Peschel] "Aus Elisée Reclus' physikalischer Erdkunde, 1, Die fliessenden Wasser," *Das Ausland*, vol. 41, no. 19 (7 May 1868), 433-438; [Peschel] "Aus Elisée Reclus' physikalischer Erdkunde, 3, Der Meeresboden," *Das Ausland*, vol. 42, no. 29 (17 July 1869), 673-678; Robert E. Dickinson, *The Makers of Modern Geography* (New York: Frederick E. Praeger, 1969), 57; Alfred Hettner, "Die Entwicklung der Geographie im 19. Jahrhundert," *Geographische Zeitschrift*, vol. 4, no. 6 (1898), 313; cf. Emmanuel de Martonne, *Traité de géographie physique* (Paris: Librairie Armand Colin, 1909), 18-19.

23. BN 22910: 373-375 (*Corr.*, I, 296-297).

24. [Oscar Peschel] "Aus Elisée Reclus' physikalischer Erdkunde, 4, Die belebte Schöpfung," *Das Ausland*, vol. 42, no. 34 (21 August 1869), 811-813.

25. George Perkins Marsh Collection, Bailey Library, University of Vermont; Bigelow, *Retrospections*, IV, 160-163, 169-170, 559-561; David Lowenthal, "George Perkins Marsh on the Nature and Purpose of Geography," *Geographical Journal*, vol. 126, part 4 (December 1960), 413-417; BN 22914: 284-286.

Apart from the loss of his papers in 1871, Reclus apparently had previously not made a systematic effort to retain copies of incoming letters. In 1864 Bigelow mentioned that Reclus had given him "some autographs . . . from his own correspondence" to be sold at a charity bazaar in New York City to aid Union soldiers and their families (*Retrospections*, II, 167).

26. Elisée Reclus, *Histoire d'un ruisseau* (Paris: Bibliothèque d'Education et de Récréation, J. Hetzel et Cie, n. d. [1869]), 1, 313-317.

27. *Ibid.*, 40, 172-181, 199, 209-211; Letter from Patrick Geddes to his wife, 9 September 1898, Geddes Papers, National Library of Scotland.

28. *Corr.*, I, 229; Nettlau, *Elisée Reclus*, 97-98; P. Reclus, *Frères*, 45; Ishill, *Reclus*, 6-7; Eugène Noel, *Fin de vie (notes et souvenirs)* (Rouen: Imprimerie Julien Lecerf, 1902), 67-68 *et passim*. Eugène Noel (1816-1899) was an intimate of Alfred Dumesnil and came to know the Reclus well. Elisée wrote the preface to Noel's posthumously published memoirs. Noel's description of the Reclus is very apt: "Nowhere is the law of atavism more marked than in that Protestant family of the Reclus. Zealous and austere, they manifest the spirit of the Pastors of the Desert, with the same inclination to saintliness. Full of illusions and hope, they have before them the mirage of the Promised Land; they run towards it and would want to see others run to it as well" (*Fin de vie*, 185).

29. Ruth Putnam, ed., *Life and Letters of Mary Putnam Jacobi* (New York: G. P. Putnam's Sons, 1925), 191-192.

30. *Ibid.*, 197.

31. *Ibid.*, 229.

32. *Ibid.*, 171-172, 177, 238.

33. BN 22909: 9; P. Reclus, *Frères*, 62, 95; Ishill, *Reclus*, 11-12.

34. Putnam, *op. cit.*, 206, 208, 238.

35. Putnam, *op. cit.*, 225; Elie Reclus, "Our Trip to Egypt as Guests of the Viceroy," *Putnam's Magazine*, vol. 5, no. 27 (March 1870), 328-343; Elie Reclus, "Voyage au Caire et dans la Haute-Egypte," *La Philosophie positive*, vol. 6, no. 5 (March-April 1870), 241-270, no. 6 (May-June 1870), 503-527, vol. 7, no. 1 (July-August 1870), 127-152; Nettlau, *Elisée Reclus*, 136; *Corr.*, I, 337-343.

36. Jean Dubois, *Le Vocabulaire politique et social en France de 1869 à 1872* (Paris: Librairie Larousse, 1962), 110, 138, 185, 281-282; P. Reclus, *Frères*, 58; Benjamin

Harrison, "Gabriel Monod and the Professionalization of History in France, 1844-1912," Unpublished Ph.D. thesis, University of Wisconsin, 1972, p. 137.

Notes to Chapter Four

1. BN 22909: 10-16
2. Ruth Putnam, op. cit., 255.
3. Ibid., 271.
4. BN 24282 (Autographes Félix et Paul Nadar): 8188 (reprinted in Corr., II, 3).
5. Jean Steens, "Profils socialistes," Le Correspondant, n. s. vol. 175, no. 6 (25 June 1903), 1177; Letters from Reclus to Nadar, November 1870, copies in Institut français d'histoire sociale, Fonds Elisée Reclus; Putnam, op. cit., 280.
6. Putnam, op. cit., 276.
7. Ibid., 277. Elie Reclus visited Mary Putnam Jacobi (She had married Abraham Jacobi, an eminent doctor who shared her advanced views) in the United States in 1877. She tried unsuccessfully to get him a writing job through her husband's friend Carl Schurz. Letter from Mary Putnam Jacobi to Carl Schurz, 25 December 1876, Mary Putnam Jacobi Collection, Radcliffe College. An excellent popular biography of Mary and Abraham Jacobi appeared in 1952 (Rhoda Truax, The Doctors Jacobi [Boston: Little, Brown and Company, 1952]). The biography was slightly fictionalized in that the author improvised conversations, but these were skillfully done, in my opinion.
8. P. Reclus, Frères, 67; Corr., II, 14; Le Révolté (Geneva), vol. 1, no. 5 (7 May 1879), 3.
9. Corr., II, 23. The date 10 August refers to 10 August 1792, the communal assault on the Tuileries and the suspension of the monarchy.
10. P. Reclus, Frères, 68; Kropotkin, "Elisée Reclus," Les Temps Nouveaux, vol. 11, no. 11 (15 July 1905), 2; "Rapport du Capitaine Commandant par interim le 119e Bon de la Garde Nationale de la Seine sur la journée sur la journée [sic] du 4 avril. Affaire de Châtillon," 4-p. Ms. (R^{c}MS 68) in Bibliothèque Municipale de Saint-Denis; (Anon.) 1871. Enquête sur la Commune de Paris, 2nd ed. (Paris: Editions de la Revue Blanche, n. d.), 53-56; Bernard Noël, Dictionnaire de la Commune (Paris: Fernand Hazan Editeur, 1971), 71-72, 276; P.-O. Lissagaray, Histoire de la Commune de 1871, 6th ed. (Paris: E. Dentu, n. d.), 185.
11. Letter from Reclus to Joukovsky, n. d. (1877), copy in Institut français d'histoire sociale, Fonds Elisée Reclus; 1871, Enquête (cited above in fn 10), 53-56.
12. Elisée Reclus, L'Evolution, la révolution et l'idéal anarchique (Paris: P.-V. Stock, 1898), 285-287.
13. Elie Reclus, La Commune de Paris au jour le jour, 1871, 19 mars—28 mai (Paris: Librairie C. Reinwald, Schleicher Frères, Editeurs, 1908), 1; Henri Dubief, "L'Administration de la Bibliothèque Nationale pendant la Commune," Le Mouvement Social, no. 37 (October-December 1961), 3-16; Jean Maitron, ed., Dictionnaire biographique du mouvement ouvrier français, vol. 8 (Paris:Les Editions Ouvriers, 1970), 299, 301.
14. Lissagaray, op. cit., 547-548; Corr. II, 29-72 passim; Institut français d'histoire sociale, Fonds Elisée Reclus, transcripts of letters 17 May 1871 to 3 October 1871; Nettlau, Elisée Reclus, 162-163.

15. A. Parménie and C. Bonnier de la Chapelle, *Histoire d'un éditeur et de ses auteurs, P.–J. Hetzel (Stahl)* (Paris: Editions Albin Michel, 1953), 556-558; Letters from Elisée Reclus to Fanny Reclus, 2 August 1871, and Lois Reclus, 23 August 1871, copies in Institut français d'histoire sociale, Fonds Elisée Reclus; *Corr.*, II, 50-52, 59-60.

16. BN 22909: 25-26, 22911: 82; Bigelow, *op. cit.*, vol. 4, 561; Ishill, *Reclus*, 109-110; cf. Nadar, *Sous l'incendie* (Paris: G. Charpentier, 1882), vii-viii.

17. "Rapport à la Commission des grâces," 16 January 1872, and letter from Ministère de la Guerre to Ministère de la Justice, 29 December 1871, in Archives Nationales, BB^{24}/732; P. Reclus, *Frères*, 81; Eugene Oswald, *Reminiscences of a Busy Life* (London: Alexander Moring, Ltd., 1911), 427-429, 511. It is not known why Oswald did not use all of the "hundred excellent signatures" on the first petition. There were actually sixty-one signatures on the petition dated 30 December 1871 and thirty-three on the petition of 20 March 1872 (by which time Reclus was a free man). See copies of *Le Courrier de l'Europe*, 20 January and 23 March 1872, in BN 22909: 36-39.

18. Bigelow, *op. cit.*, vol. 5, 10, 16; letter from the Ministère de la Guerre to Ministère de la Justice, 8 February 1872, Archives Nationales, BB^{24}/732; P. Reclus, *Frères*, 85; Note, "La Veille de la Libération," 14 March 1872, copy in Institut français d'histoire sociale, Fonds Elisée Reclus.

Notes to Chapter Five

1. Letter from Elisée Reclus to John Bigelow, 19 April 1872, cited in Bigelow, *Retrospections*, V, 26-27. Also see Chapter 3, Note 15 (above). According to my calculations, 125 German thalers would have been the equivalent of about ninety American dollars in 1872. In a letter to Oswald on 21 March 1872, before he left Zurich, Reclus cited the advantages of Lugano as being "the climate of Italy, Swiss liberty, and the nearness of Vienna, where geographical and geological documents are found in such great quantity" (*Corr.*, II, 93). The editor suggested that Vienna might have been a slip of the pen and that Milan was intended. In any event, the isolation of Lugano soon became apparent to Reclus.

2. Letter from Michael Bakunin to Elisée Reclus, 15 February 1875, in James Guillaume, *L'Internationale, Documents et souvenirs (1864-1878)*, III (Paris: P.–V. Stock, 1909), 284; Carl Landauer, *European Socialism* (Berkeley and Los Angeles: University of California Press, 1959), I, 126; Erich Gruner, "La Suisse et le tournant historique de 1870-1871," *Revue d'histoire moderne et contemporaine*, vol. 19, no. 2 (April-June 1972), 243-245.

3. Arthur Lehning, ed., *Michel Bakounine et l'Italie, 1871-1872* (Leiden: E. J. Brill, 1961), 248; Ishill, *Reclus*, 200-201; P. Reclus, *Frères*, 60-61.

4. Ishill, *Reclus*, 202, 207-208; Landauer, *op. cit.*, I, 128; Guillaume, *op. cit.*, IV, 36-37; Charles Thomann, *Le Mouvement anarchiste dans les montagnes neuchâteloises et le Jura bernois* (La Chaux-de-Fonds: Imprimerie des Coopératives Réunies, 1947), 90. Thomann then added (p. 91) an interesting story which is very revealing of Reclus's habits and those of his anarchist friends. Joseph Favre described an anarchist meeting at Lugano at which Elisée Reclus drank water, Malon and Arnould drank red wine, Malatesta, Guesde, and Favre drank white wine, and Bakunin, after having a glass of

beer, drank several glasses of tea while filling the air with clouds of smoke, all of which was disagreeable to Reclus.

5. There is a large literature on the decline of French science after the 1840s and the alleged causes and consequences; see, for example, Joseph Ben-David, "The Rise and Decline of France as a Scientific Centre," *Minerva*, vol. 8, no. 2 (April 1970), 160-179; Robert Fox, "Scientific Enterprise and the Patronage of Research in France 1800-70," *Ibid.*, vol. 11, no. 4 (October 1973), 442-473; Harry W. Paul, *The Sorcerer's Apprentice: The French Scientists' Image of German Science, 1840-1919* (University of Florida, Social Science Monograph No. 44, 1972); and Gabriel Petit and Maurice Leudet, *Les Allemands et la science* (Paris: Librairie Félix Alcan, 1916). A recent article of particular importance is Numa Broc, "La Géographie française face à la science allemande (1870-1914)," *Annales de géographie*, vol. 86, no. 473 (January-February 1977), 71-94. For Reclus's views, see Université Nouvelle de Bruxelles, *Séance solennelle de rentrée du 22 octobre 1895: Discours de MM. Elisée Reclus, Camille Moreau et Paul Janson* (Brussels: Imprimerie Veuve Ferdinand Larcier, 1895), 9-13, and *L'Evolution, la Révolution et l'idéal anarchique* (Paris: Librairie P.–V. Stock, 1898), 115-116. Reclus's proposal to Penck was mentioned in his letter to Charles Perron, 28 December 1891 (*Corr.*, III, 100-101).

6. *La République Française*, 2nd year, no. 100 (15 February 1872), 2; no. 260 (26 July 1872), 1-2; no. 330 (4 October 1872), 1-2; 4th year, no. 950 (19 June 1874), 1; Theodore Zeldin, *France 1848-1945*, vol. 1, *Ambition, Love and Politics* (Oxford: Clarendon Press, 1973), 618-619. Reclus was credited by his sister Louise with having written twenty-five unsigned articles in *La République Française* from 15 February 1872 to 8 January 1875. All but one of the articles bore the title, "Géographie générale," and the other, "Géographie polaire," was actually written, according to Reclus, by his cousin, Franz Schrader. Max Nettlau thought that he detected a definite anti-German bias in Reclus's *République Française* articles. He said that even Reclus was influenced by the "romantic-nationalistic tradition," which considers "everything on the German side reactionary and everything on the Latin, Slav, and Hungarian side as freedom-loving." See Nettlau, *Elisée Reclus*, 176.

7. Raoul Frary, *La Question du latin* (Paris: Librairie Léopold Cerf, 1885), Chapter 15, "La Géographie," 247-283. Frary was concerned with the reform of secondary education, and he discussed the disciplines that would be expanded to take the place of Latin and Greek, which would be relegated to seminaries. He gave his optimistic view that "geography . . . appears to us as the master branch of secondary teaching. It is she who will inherit the greatest part of the time and work which the abandonment of the dead languages would leave vacant" (pp. 282-283). For Vidal's life and work, see Philippe Pinchemel, "Paul Vidal de la Blache," *Geographisches Taschenbuch 1970/72* (Wiesbaden, 1972), 266-279. The development of the Vidalian school of geography is best treated in Vincent Berdoulay's Ph.D. thesis, "The Emergence of the French School of Geography (1870-1914)" (University of California, Berkeley, 1974).

8. "Plan de géographie descriptive fourni à Templier," March 1872, Institut français d'histoire sociale, Fonds Elisée Reclus (14/AS/232), Dossier III.

9. Letter from Emile Templier to Elisée Reclus, 27 May 1872, *Ibid.*

10. Archives Librairie Hachette, Dossier on Elisée Reclus.

11. Templier to Reclus, 31 August 1872, IFHS, Fonds Elisée Reclus, Dossier III.

12. Reclus to Templier, 4 September 1872, and Templier to Reclus, 7 September 1872, *Ibid.*

13. Templier to Reclus, 22 December 1872 and 27 January 1873, *Ibid.*

14. Templier to Reclus, 16 June 1873, *Ibid.*

15. BN 22909: 18, 48-49; P. Reclus, *Frères*, 87-88, 95; Ishill, *Reclus*, 12; Nettlau, *Elisée Reclus*, 177; *Corr.*, II, 105, 145-148, 157.

16. *Corr.*, II, 150-152; P. Reclus, *Frères*, 95-96; Letter from Elisée Reclus to Edouard Bouny, 24 April 1874, IFHS, Fonds Elisée Reclus; Marriage contract between Elisée Reclus and Ermance Gonini, 13 October 1875, photocopies in possession of Mme. L. Rapacka.

17. "Livre de compte à partir du 1er Septembre 1875," IFHS, Fonds Elisée Reclus, Dossier IV. The total payment for the *NGU* actually amounted to 553,862 Fr 92 centimes.

18. Elisée Reclus, "On Vegetarianism," *The Humane Review*, vol. 1, no. 5 (January 1901), 316-324; Ishill, *Reclus*, 5-6, 14; Archives Préfecture de Police, Paris, Ba/1.237, clipping from *L'Intransigeant* (6 July 1905), "L'Anarchiste modèle," by Henri Rochefort; Luigi Galleani, "Impressioni e Ricordi su Eliseo Reclus," in Elisée Reclus, *Scritti Sociali* (Buenos Aires: I Libri di Anarchia, 1930), I, 8.

19. *Bibliographie de la France*, 2nd series, vol. 64, no. 20 (15 May 1875), 275; Letter from Templier to Reclus, 18 August 1874, IFHS, Fonds Elisée Reclus, Dossier III.

20. Templier to Reclus, 1 February 1875, 6 February 1875, and 27 February 1875, IFHS, Fonds Elisée Reclus, Dossier III.

21. Templier to Reclus, 20 July 1875 and 9 May 1876, BN 22914: 394-395, 415-416.

22. Templier to Reclus, 26 July 1876, *Ibid.*, 421-422.

23. Elisée Reclus, *Nouvelle géographie universelle*, vol. 1 (Paris: Hachette, 1876), iv; Nettlau, *Elisée Reclus*, 192n.

24. *L'Année géographique*, vol. 13 (14th year, 1875—published 1876), 407.

25. *Revue critique d'histoire et de littérature*, 9th year, 1st semester, no. 23 (5 June 1875), 361-362.

26. *Loc. cit.*

27. H. Gaidoz, "Bulletin géographique," *La Revue politique et littéraire*, 2nd series, vol. 5, no. 2 (10 July 1875), 39-43; vol. 6, no. 2 (8 July 1876), 35-39; and vol. 6, no. 25 (16 December 1876), 591-593; *République Française*, 15 November 1872; *Revue des deux mondes*, vol. 61 (1 January 1866), 266-267; *NGU*, I, 231, 239.

28. *Corr.*, II, 295.

29. *La Revue politique et littéraire*, vol. 6 (16 December 1876), 592; BN 22914: 404-408; Nettlau, *Elisée Reclus*, 181; *NGU*, II, 2, 4-5, 46-47; *L'Homme et la terre*, V, 239-240, 408-410; Adolphe Joanne, *Dictionnaire des communes de la France* (Paris: Hachette, 1864), I, xvii.

30. George Woodcock and Ivan Avakumović, *The Anarchist Prince: A Biographical Study of Peter Kropotkin* (London and New York: T. V. Boardman & Co., Ltd., 1950), 90-91.

31. Nettlau, *Elisée Reclus*, 193-194. In his recent book, *Kropotkin* (Chicago: University of Chicago Press, 1976), Martin Miller appears to have misconstrued the sense of Kropotkin's account. Miller said (p. 136), "His first encounter with Elisée Reclus turned into an argument, and though he later changed his mind Kropotkin doubted whether Reclus was a 'true socialist.'" However, Kropotkin clearly stated that the meeting was a friendly one and that his preconception of Reclus was altered at that time and not later.

32. P. Reclus, *Frères*, 104; Max Nettlau, "Elisée Reclus' Briefwechsel," *Archiv für die Geschichte des Sozialismus und der Arbeiterbewegung*, vol. 3 (1913), 512-513.

33. Letter from Max Nettlau to Joseph Ishill, 19 January 1924, Ishill Collection, Harvard University; Max Nettlau, "Peter Kropotkin at Work," in *Peter Kropotkin: The Rebel, Thinker and Humanitarian*, ed. by Joseph Ishill (Berkeley Heights, New Jersey: The Free Spirit Press, 1923), 14.

34. G. D. H. Cole, *Socialist Thought: Marxism and Anarchism, 1850-1890* (London: Macmillan & Co., Ltd., 1957), 345; *Le Révolté*, vol. 1, no. 1 (22 February 1879), 1; Miller, *op. cit.*, 167; Jean Maitron, *Histoire du mouvement anarchiste en France (1880-1914)* (Paris: Société universitaire d'éditions et de librairie, 1951), 563-587 *passim*; Jacques Prolo, *Les Anarchistes* (Paris: Librairie des sciences politiques et sociales, Marcel Rivière et Cie., 1912), 77; Jean Grave, *Le Mouvement libertaire sous la 3ᵉ république (Souvenirs d'un révolté)* (Paris: Les Oeuvres Représentatives, 1930), 147, 294.

35. [Elisée Reclus] "Les Produits de l'industrie," *Le Révolté*, 3-part article: 8th year, nos. 45, 47, and 49 (February-April 1887); "La Richesse et la misère," *Le Révolté/La Révolte*, 6-part article: *Le Révolté*, 9th year, no. 12, to *La Révolte*, 1st year, no. 8 (June-November 1887); Letter from Elisée Reclus to Richard Heath, 1884, *Corr.*, II, 325; Elisée Reclus, review of René Gonnard, *La Dépopulation en France* (1898), *L'Humanité nouvelle*, vol. 4, no. 21 (10 March 1899), 377; Elisée Reclus, review of Gabriel Giroud, *Population et subsistances, Essai d'arithmétique économique* (1904), *La Revue*, vol. 51 (1 September 1904), 100-101; Jean Maitron, *Le Mouvement anarchiste en France*, I (Paris: Maspero, 1975), 344-349; cf. Kropotkin's views of population as reported by James Hulse, *Revolutionists in London: A Study of Five Unorthodox Socialists* (Oxford: Clarendon Press, 1970), 62-63. Reclus's expressions of the Bounty of Earth have an almost Biblical ring, and Béatrice Giblin sees them as additional marks of his Protestant upbringing. See B. Giblin, "Elisée Reclus: géographie, anarchisme," *Hérodote*, no. 2 (April-June 1976), 36.

36. *Corr.*, II, 266n, 268-270; Nettlau, *Elisée Reclus*, 229; Maitron, *Histoire. . . . ,op. cit.*, 156; Miller, *op. cit.*, 164.

37. For information about Gustave Lefrançois *dit* Lefrançais (1826-1901), see Noël, *Dictionnaire*, 233-234; P. Reclus, *Frères*, 97; and Jean Maitron, ed., *Dictionnaire biographique du mouvement ouvrier français*, part 2, vol. 7 (Paris: Les Editions ouvrières, 1970), 90-93. For Léon Metchnikoff (1838-1888), see especially Elisée Reclus's "Préface" (pp. v-xxviii) to Metchnikoff's *La Civilisation et les grands fleuves historiques* (Paris: Hachette, 1889). For Sensine, see Nettlau, *Elisée Reclus*, 232n, and P. Reclus, *Frères*, 97. Sensine's extravagant praise of Reclus was printed in his *Chrestomathie française du XIXe siècle*, 6th ed. (Lausanne: Librairie Payot, 1914), I, 581.

38. Letter from Paul Reclus to Max Nettlau, 27 June 1924, in Ishill Collection, Harvard University.

39. Elisée Reclus, "Attila de Gerando," *Revue de géographie*, vol. 42, no. 1 (January 1898), 1-4; *Corr.*, II, 135 *et passim*; and Nettlau, *Elisée Reclus*, 176. The quotations are from letters from Élisée Reclus to Gerando, 25 June 1877 and n. d., IFHS, Fonds Elisée Reclus.

40. *Corr.*, II, 162; Letter from Reclus to Gerando, 9 December 1882, IFHS, Fonds Elisée Reclus.

41. Letter from Reclus to Gerando, 26 September 1879, IFHS, Fonds Elisée Reclus.

42. *Corr.*, II, 239-240.

43. H. Roorda van Eysinga, "Elisée Reclus propagandiste," *La Société nouvelle*, 2nd series, 13th year, no. 2 (August 1907), 186.

44. On the matter of amnesty, see especially Jean Joughin, *The Paris Commune in French Politics, 1871-1880: The History of the Amnesty of 1880*, 2 vols. (continuously paged) (The Johns Hopkins University Studies in Historical and Political Science, series 73, nos. 1-2, 1955) (Baltimore: The Johns Hopkins Press, 1955), 85, 111, 196, 226, 237-238, 465.

For the wedding of Reclus's daughters, see letter from Reclus to Nadar, 11 October 1882, IFHS, Fonds Elisée Reclus; *Corr.*, II, 265 *et passim*; Nettlau, *Elisée Reclus*, 223-224; P. Reclus, *Frères*, 108; *Le Révolté*, 4th year, no. 19 (11 November 1882), 3. Reclus's remark about opposition to the "juridical family" was made in his paper, "Evolution et révolution," *Le Révolté*, 1st year, no. 27 (21 February 1880), 3.

45. Ishill, *Reclus*, 108; P. Reclus, *Frères*, 110-111, 118, 123; Nettlau, *Elisée Reclus*, 235, 257.

46. Marian Brun, "Département des estampes et cartes de la Bibliothèque publique et universitaire de Genève," Vereinigung Schweizerischer Bibliothekare, *Nachrichten*, Year 41 (1965), no. 4, 113-114; Letter to Elisée Reclus from the President of the Administrative Council, City of Geneva, 17 May 1890, in possession of Mme. L. Rapacka; Letters from Charles Knapp to Ermance Reclus, 10 October 1890, and to Elisée Reclus, 12 February 1897, in Archives of the Société neuchâteloise de géographie.

47. Ishill, *Reclus*, 39-40, 113-114; J. Maitron, *Histoire*, *op. cit.*, 175-177; Elisée Reclus, "L'Evolution de la morale. Le Vol et les voleurs," *La Révolte*, 2nd year, no. 22 (10-16 February 1889), 1-2; Maitron, *Mouvement*, *op. cit.*, I, 191-193.

48. *Corr.*, III, 235; Mermeix (pseudonym of Gabriel Terrail), *La France socialiste: Notes d'histoire contemporaine* (Paris: F. Fetscherin et Chuit, Editeurs, 1886), 213; Zeldin, *op. cit.*, 778; Nettlau, *Elisée Reclus*, 271.

49. *Corr.*, II, 450-482 *passim*; NGU, XVI, 706-708, 743, 756; cf. Elisée Reclus, "La Grève d'Amérique," *Le Travailleur*, 1st year, no. 5 (September 1877), 6-16, and R. Laurence Moore, *European Socialists and the American Promised Land* (New York: Oxford University Press, 1970).

50. NGU, XIX, 793-796; Elisée Reclus, *Nouvelle géographie universelle: La Terre et les hommes: Tableaux statistiques de tous les états comparés: Années 1890 à 1893* (Paris: Hachette, 1894).

51. Altogether, the NGU comprised 16,983 pages (average of 894 per volume) and included 3,508 maps and 1,407 gravures. See also "L. G." (Lucien Gallois), review of NGU, Vol. 19, in *Annales de géographie, Bibliographie de l'année 1894* (Paris: Armand Colin, 1895), 236-237; C. Knapp, "Elisée Reclus," *Bulletin de la Société neuchâteloise de géographie*, vol. 16 (1905), 313.

52. Geoffrey Martin, *Ellsworth Huntington: His Life and Thought* (Hamden, Connecticut: Archon Books, 1973), 17; Lucien Febvre, *La Terre et l'évolution humaine* (Paris: La Renaissance du Livre, 1922), 19.

53. P. Vidal de la Blache and L. Gallois, eds., *Géographie universelle*, 15 vols. (Paris: Armand Colin, 1927-1946); P. R. Crowe, "The New 'Géographie Universelle': A Review," *Scottish Geographical Magazine*, vol. 44, no. 1 (January 1928), 42: O. H. K. Spate, *Let Me Enjoy* (London: Methuen, 1966), 112. The phrase "bibliographic dinosaurs" seems to have been coined by J. K. Wright in his paper, "Some British 'Grandfathers' of American Geography," in *Geographical Essays in Memory of Alan G. Ogilvie*, ed. by R. Miller and J. W. Watson (London, etc.: Thomas Nelson and Sons, Ltd., 1959), 147.

54. BN 22915: 57-61.

Notes to Chapter Six

1. Raymond Jacqmot, "L'Affaire Elisée Reclus, ou les effets d'une bombe!," *Bulletin mensuel de l'Union des anciens étudiants de l'Université libre de Bruxelles,* vol. 31, no. 257 (April 1958), 8.
The full story of the controversies in the Free University of Brussels in the early 1890s is very complex and will not be repeated here, but the interested reader is referred to Jacqmot's article in the *Bulletin* (above), vol. 31, no. 257 (April 1958), 5-13, and no. 258 (May 1958), 11-16; and especially to E.-F.-A. Goblet d'Alviella, *L'Université de Bruxelles pendant son troisième quart de siècle, 1884-1909* (Brussels: M. Weissenbruch, 1909), 26-37.

2. A good sketch of the New University is given by Andrée Despy-Meyer in the opening pages of her *Inventaire des archives de l'Université Nouvelle de Bruxelles (1894-1919)* (Brussels: Université Libre de Bruxelles, Service des Archives, 1973). See also A. Despy-Meyer and Pierre Goffin, *Liber Memorialis de l'Institut des hautes études de Belgique fondé en 1894* (Brussels: Université Libre de Bruxelles, Service des Archives, and Institut des hautes études de Belgique, 1976).

3. Jacqmot, *op. cit.,* 15; Edmond Picard, "L'Institut des hautes études à l'Université Nouvelle de Bruxelles," *L'Humanité nouvelle,* vol. 1 (1897), 557.

4. Ishill, *Reclus,* 13-14. In the original French version of Paul Reclus's essay for the Ishill volume, used by Nettlau for his biography of Elisée Reclus (Nettlau, *Elisée Reclus,* 292), it said that Reclus died in Mme. de Brouckère's arms, but this phrase was deleted in the English version, perhaps by Ishill or his wife in the fear that the reader might assume a romantic involvement.
Ermance Reclus was described as an intelligent woman, "a remarkable botanist and entomologist, a woman of very delicate and balanced knowledge and charm which often contrasted with the youthful impetuosity of her husband" (Guillaume De Greef, *Eloges d'Elisée Reclus* . . . [Ghent, 1906], 25). She published a charming little children's book in 1895 which contains stories of her own childhood sixty years earlier (*Vacances chez le grand-père* [Paris: Hachette]). She died in 1918.

5. Most of the lecture outlines in Reclus's course in comparative geography, "Géographie comparée dans l'espace at dans le temps," were printed and can be consulted in the New University archives (1Z456-458) which have been deposited in the archives of the Free University of Brussels. In the first five years of teaching this course (1894-1899), Reclus covered East, South, and Southwest Asia, including the Caucasus and Egypt.

6. Despy-Meyer, *op. cit.,* 8; Guillaume De Greef, "Discours de M. Guillaume De Greef . . . 17 octobre 1898," 12-page supplement to *L'Humanité nouvelle,* vol. 3 (1898); *Le Révolté,* no. 26 (7 February 1880), 4.

7. P. Reclus, *Frères,* 142-143; *Corr.,* III, 172-173, 185-187, 210.

8. Elisée Reclus, "L'Enseignement de la géographie," reprint from no. 1 (1903) of *Bulletin de la Société belge d'astronomie,* p. 5; Reclus, "Art and the People," in Ishill, *Reclus,* 329-330; *Nouvelle géographie universelle,* III, 123; cf. René Chaughi, "Les voyages," *Les Temps nouveaux,* 6th year, no. 39 (19-25 January 1901), 1; G. S. Dunbar, "Why Travel?," *Landscape,* vol. 21, no. 3 (Spring-Summer 1977), 45-46.

9. Edmond Picard, "L'Institut des hautes études à l'Université nouvelle de Bruxelles," *L'Humanité nouvelle,* vol. 1 (1897), 557-558; Ishill, *Reclus,* 229.
In 1874 Auguste Baud-Bovy had asked Reclus to give a series of lectures on "Geography As Applied to History" in Geneva. Reclus accepted but told Baud-Bovy

of his deficiencies as an orator. "You ask if I have ever lectured before a large audience. I began with a fiasco before an audience of distinguished personages and bourgeois in a hall on the Rue des Capucines. This fiasco was followed by two demi-successes in the same hall and before the same audience. Later, during the Siege, at the town-hall of the 3rd arrondissement I gave a series of lectures before 150 or 200 schoolteachers who were very pleased." Letter from Reclus to Baud-Bovy, 21 (?) November 1874, Baud-Bovy Collection, Département des manuscrits, Bibliothèque publique at universitaire de Genève.

The "fiasco" was undated, but it might have been the lecture which Reclus mentioned in a letter to his mother in 1861: "Recently I had occasion to speak on the sea and its currents before a rather large audience which included Messrs. Michelet, Carnot, Legouvé, J. Simon and Pelletan. Unfortunately, I was much more frightened than I expected to be and I floundered a little: I hope to have the occasion soon to redeem myself" (Corr., I, 215).

10. Jacques Lavalleye, "Jean-Jacques Mesnil," Académie Royale des Sciences, des Lettres et des Beaux-Arts de Belgique, Biographie Nationale, vol. 34, Supplément, vol. 6, no. 2 (1968), cols. 596-605; Jacques Mesnil, "Elisée Reclus," Monde (15 March 1930); Clara Mesnil, "Souvenirs sur Elisée Reclus," Ibid., expanded as "Quelques souvenirs sur Elisée Reclus," Maintenant, no. 2 (1946), 252-256; Jacques Mesnil, "Biographie d'Elisée Reclus," unpublished 50-page typescript, copies in Académie Royale de Belgique and Internationaal Instituut voor Sociale Geschiedenis (Archief Elisée Reclus); Jacques Mesnil, "Baedeker," Europe, Revue mensuelle, vol. 15, no. 59 (15 November 1927), 414-416, translated and edited by G. S. Dunbar, "The Way It Was Done in Leipzig: A Comment on Baedeker's First Century," Landscape, vol. 19, no. 3 (May 1975), 11-13.

11. Mesnil, Eliseo Reclus (Mantova, 1905), 24; Ishill, Reclus, 189.

12. Mesnil, Eliseo Reclus, 21.

13. Emile Vandervelde, Souvenirs d'un militant socialiste (Paris: Les Editions Denoël, 1939), 37.

14. For Wyld's globe, see chapter 2 (above). The discussion of Reclus's Great Globe scheme is condensed from G. S. Dunbar, "Elisée Reclus and the Great Globe," Scottish Geographical Magazine, vol. 90, no. 1 (April 1974), 57-66, to which the reader can go for full documentation. Only material not used in the article will be included in the notes here.

15. "Ch.D." (Charles Delfosse), "Elisée Reclus, géographe," Les Hommes du jour, 1st series, no. 32a (n. d. [1895-1896]); "G.D." (Dallet), Annales de géographie, Bibliographie de l'année 1895 (1896), 20; "E.F." (identity unknown), "Le Globe terrestre de 1900," L'Illustration, vol. 111, no. 2871 (5 March 1898), 183.

16. Despy-Meyer, op. cit., 10-12; Le Moniteur international (Liége), vol. 1, no. 5 (7 August 1898), 1-3; Letters from Reclus to Ostroga in possession of Mme. Rapacka; Letters from Reclus to Ostroga and Régnier in Institut français d'histoire sociale, Fonds Elisée Reclus; P. Reclus, Frères, 147; Manuscript minutes of Council of the Société anonyme, 10 October 1899, archives of the Université Nouvelle (1Z419) in Université Libre archives.

17. P. Reclus, Frères, 146; Nettlau, Elisée Reclus, 315; Ishill, Reclus, 16; Letter to the author from A. De Smet, 19 October 1971.

18. For Engerrand, see especially the Bulletin of the Texas Archeological Society, vol. 32 (1961), "The George C. Engerrand Volume," which contains the following articles: A. P. Brogan, J. G. McAllister, and T. N. Campbell, "In Memoriam: George

Charles Marius Engerrand," 1-8; W. W. Newcomb, Jr., "George C. Engerrand in Europe, 1898-1907," 9-17; and John A. Graham, "George C. Engerrand in Mexico, 1907-1917," 19-31. The first article was also published under T. N. Campbell's name alone in *American Anthropologist*, vol. 64, no. 5, part 1 (October 1962), 1052-1056. The quotations from letters from Elisée Reclus (either directly or through Louise Dumesnil or Magali Cuisinier) to Engerrand are from the Engerrand Papers which were loaned to the author by Engerrand's daughter, Mrs. Anita Gafford of Denton, Texas.

19. To my knowledge no complete list of the *Publications* of the Institut géographique de Bruxelles is available in any library catalogue or publication, and so for that reason I am giving such a list here:

1. B. Joseph de Siemiradzki. "La Nouvelle Pologne, Etat de Paraná (Brésil)." 1899. 11 pages.
2. Vaughan Cornish. "Formation des dunes de sable." 1900. 37 p. Translated by Emile Cammaerts from Cornish's article, "On the Formation of Sand-Dunes," *Geographical Journal*, vol. 9, no. 3 (March 1897), 278-302.
3. Henry Benest. "Fleuves sous-marins. Epanchements d'eaux douces au-dessous du niveau de la mer." 1900. 31 p. A modified translation of Benest's article, "Submarine Gullies, River Outlets, and Fresh-Water Escapes beneath the Sea-Level," *Geographical Journal*, vol. 14, no. 4 (October 1899), 394-413.
4. Valère Maes. "Projection sphérique comparée aux autres projections." 1901. 13 p.
5. Elisée Reclus. "L'Enseignement de la géographie. Globes, disques globulaires et reliefs." 1901. 10 p.
6. G. Marinelli. "L'Accroissement du delta du Po au XIXeme siécle." 1901. 36 p. Translation by George Engerrand (?) of Marinelli's article, "L'Accrescimento del delta del Pol nel secolo XIX," *Rivista geografica italiana*, vol. 5 (1898), 24-37, 65-85.
7. G. Guyou (Paul Reclus). "Un nouveau planétaire." 1902. 13 p.
8. M. R.-H. Codrington. "La Magie chez les insulaires mélanésiens." 1903. 31 p. Translated by Emile Cammaerts from Robert Henry Codrington's book, *The Melanesians: Studies in Their Anthropology and Folk-Lore* (Oxford: Clarendon Press, 1891).
9. Pierre Kropotkine (Peter Kropotkin). "Orographie de la Sibérie avec un aperçu de l'orographie de l'Asie." 1904. 119 p. Translated from various works by Kropotkin, especially "The Orography of Asia," *Geographical Journal*, vol. 23, nos. 2 & 3 (February and March 1904).
10. Edgar Sacré. "L'Esperanto. Langue internationale auxiliaire." 1905. 15 p. (Note—Esperanto was of great interest to anarchists because of its "supranational character" [Jean Chesneaux, "Critique sociale et thèmes anarchistes chez Jules Verne," *Le Mouvement social*, no. 56 (July-September 1966), 62].)

20. Ishill, *Reclus*, 16-17.
21. Elisée Reclus, *L'Homme et la terre*, vol. 1 (Paris: Librairie Universelle, 1905), i-ii.
22. G. S. Dunbar, "Some Early Occurrences of the Term 'Social Geography,' " *Scottish Geographical Magazine*, vol. 93, no. 1 (April 1977), 15-20; G. S. Tikhomirov,

"Elize Rekliu—vydaiushchiisia frantsuzskii geograf," *Istoriia geologo-geograficheskikh nauk*, no. 2 (Trudy Instituta Istorii Estestvoznaniia i Tekhniki, vol. 37) (Moscow, 1961), 49.

23. Letters from Hachette to Elisée Reclus, 3 October and 1 December 1894, Archives Librairie Hachette.

24. Jean Mistler, *La Librairie Hachette de 1826 à nos jours* (Paris: Hachette, 1964), 264-265.

25. *Ibid.*, 265.

26. *Corr.*, III, 184-185.

27. *Ibid.*, 260-262; Nettlau, *Elisée Reclus*, 284, 316.

28. Nettlau, *Elisée Reclus*, 317-318.

29. Ludmila Vachtová, *Frank Kupka, Pioneer of Abstract Art* (New York: McGraw-Hill Book Company, 1968), 23-25, 41-43, 253; Denise Fédit, *L'Oeuvre de Kupka* (Paris: Editions des Musées Nationaux, 1966), Section "Oeuvres figuratives de 1900 à 1911," unpaged. Kupka's original drawings for *L'Homme* are now in the National Gallery of Prague.

30. Nettlau, *Elisée Reclus*, 318n; *L'Homme et la terre*, vol. 6 (1908), 543.

31. References to Marx in *L'Homme*: vol. 5, 231; vol. 6, 336. Reclus's most extended discussion of Marx was in his preface to F. Domela Nieuwenhuis's *Le Socialisme en danger* (Paris: P.-V. Stock, Editeur, 1897). Reclus said that Marx was unfortunately idolized and that the German Socialist politicians were twisting his words. As in the case of Jesus Christ, Marx's followers were paying reverent homage to their master while denying him in practice. The danger of socialism, according to Reclus, lies in the misguided belief of Marx's disciples that it is necessary to gain votes and get power and then, to stay in power, it is necessary to use force and maintain police and armies. Reclus did not totally oppose Marx or use invective against him, as Bakunin and Kropotkin did; he simply thought that Marx was misguided, even though he shared most of the same goals. Not surprisingly, many Marxists have the same feelings about Reclus.

32. Elisée Reclus, "The Evolution of Cities," *Contemporary Review*, vol. 67, no. 2 (February 1895), 250-251; *L'Homme et la terre*, V, 341-342. M. Gaubert de Ger was the inventor of a machine which would compose and distribute type. The machine, named the *Gérotype* after his birthplace, was described in several *Comptes rendus hebdomadaires des séances de l'Académie des sciences* (10 May 1841-26 December 1842) and in a 15-page pamphlet published by Gaubert (*Notice sur le Gérotype . . . Rapport à l'Académie des sciences, le 5 décembre 1842* [Paris, 1843]), but none of these sources mentions the subject of the spacing of towns. It is possible that Reclus encountered Gaubert or his writings in England in the 1850's. I should hazard the guess that Reclus was recalling Gaubert and his publication after four decades or more and that he did not attempt to get a fresh citation. F. Thibodeau mentioned Gaubert's *Gérautype* (sic) and then cited a composing machine produced by Gobert (sic) in London in 1855. See *La Lettre d' Imprimerie*, vol. 2 (Paris: AuBureau de l'Edition, n.d. [1921]), 512. I hope to learn more about M. Gaubert and his views on the spacing of cities.

33. Reclus, "The Evolution of Cities," 257.

34. *L'Homme et la terre*, VI, 256, 311, 368.

35. *Ibid.*, 385, 50lff.

36. Letter from Paul Reclus to Mary Putnam Jacobi, 12 March 1904, Mary Putnam Jacobi Collection, Radcliffe College; [Elisée Reclus] *Elie Reclus, 1827-1904* (Paris: L'Emancipatrice (imprimerie communiste), n. d. [1905]).

37. Elie Faure, *Oeuvres complètes*, Vol. 3, *Oeuvres diverses* (Paris: Jean-Jacques Pauvert Editeurs, 1964), 767.

38. *Loc. cit.*

39. *Ibid.*, 735, 767, 1136. On the millennialism and unquenchable optimism of Reclus, Kropotkin, and the other anarchists, see also Barbara Tuchman, *The Proud Tower* (New York: Macmillan, 1966), 81, 85, 87-88.

40. Letter from Elisée Reclus to Nadar, 28 June 1904, BN 24282.

41. In the copy of the *Introduction* which is found in the Hôtel de Ville in Sainte-Foy-la-Grande, someone wrote the following note on a piece of paper glued to the title page: "Elisée Reclus pages I à XXXVIII et XLVII à LIV Paul Reclus, fils d'Elie pages XXXVIII à XLVII et LIV à CLXIII." This information cannot be found elsewhere.

42. Elisée Reclus, "Le Mont Etna et l'éruption de 1865," *Revue des Deux Mondes*, vol. 58 (1 July 1865), 138.

43. *La Terre*, I, 657; G. Bonvalot, "De Paris au Tonkin à travers le Tibet inconnu," *Le Tour du monde*, vol. 62 (2nd semester 1891), 334. This placename has not survived, but a peninsula in Antarctica which was named for Reclus by de Gerlache in his "Belgica" expedition of 1897-1899 still persists on modern maps. Guillaume De Greef, *Eloges d'Elisée Reclus . . .*, 1906, 52-53.

In 1907 the Municipal Council of Paris wanted to name a street after Elisée Reclus, but this gesture was regarded by Paul Reclus as an attempt by the bourgeoisie to subvert the memory of the old anarchist. "It could as well be called the Street of the Immaculate Conception," said Paul. Today there is an Avenue Elisée Reclus, only two blocks long, very near the Eiffel Tower in the 7th Arrondissement. Toward the end of Elisée's life, Paul told him that it was inevitable that someone would try to erect a bust in his honor in Sainte-Foy, and the old sage replied, "Well, I hope that a comrade will knock it down and put a fruit tree in its place." See Paul Reclus, "Pour Elisée Reclus," *Les Temps nouveaux*, 13th year, no. 2 (11 May 1907), 2. Although no bust or statue of Elisée Reclus was erected in Sainte-Foy, a street was named for him in 1908 and then was renamed Rue des frères Reclus in 1936, to honor not only Elisée but his four brothers as well. See Jean Corriger, "Variations des noms des rues de Sainte-Foy-la-Grande," *Sainte-Foy-la-Grande et ses alentours* (Fédération historique du Sud-Ouest, Actes du XIXe Congrès d'études régionales, 1966) (Bordeaux: Editions Bière, 1968), 195-197.

44. Elisée Reclus, "Proposition de dresser une carte authentique des volcans," Extrait du n° 11 (1903) du *Bulletin de la Société belge d'astronomie*; Elisée Reclus, *Les Volcans de la terre*, 3 fascicules (Brussels: Société belge d'astronomie, de météorologie et de physique du globe, 1906-1910); Louis Raveneau, *Annales de géographie*, XXe *Bibliographie géographique annuelle, 1910*, 45; Emm. de Margerie, "Le Géologie," *La Science française*, vol. 1 (Paris: Ministère de l'instruction publique et des beaux-arts, 1915), 219.

45. *Corr.*, III, 302-308; Nettlau, *Elisée Reclus*, 332-333; Letter from Paul Reclus to Mary Putnam Jacobi (cited above in fn. 36).

46. *Corr.*, III, 326-327; De Greef, *Eloges*, 47.

Notes to Chapter Seven

1. Biographical data on Elisée Reclus's brothers and sisters are scattered throughout numerous sources, especially as incidental information in works that are devoted to

their more famous brother. See especially Paul Reclus, "A Few Recollections on the Brothers Elie and Elisée Reclus," in Ishill, *Reclus*, 9-10. Perceptive contemporary descriptions of the five brothers are given by Gabriel Astruc, "La Famille Reclus," *L'Illustration*, vol. 102, no. 2652 (23 December 1893), 599; and R. Marzac, "Les Reclus," *Le Figaro*, 40th year, 3rd series, no. 211 (30 July 1894), 1-2. Emile Zola read the *Figaro* article and used the Reclus as models for some characters in his novel *Paris* (1898). See René Ternois, *Zola et son temps: Lourdes-Rome-Paris* (Publications de l'Université de Dijon, 22) (Paris: Société Les Belles Lettres, 1961), 348n. For insightful obituaries of Onésime and Paul (the doctor), see Dr. Montgré, "Paul Reclus," *Larousse mensuel illustré*, vol. 3, no. 100 (June 1915), 464, and Henri Froidevaux, "Onésime Reclus," *Ibid.*, vol. 3, no. 117 (November 1916), 930-931. Another posthumous appreciation of Paul was given by Léon Daudet in *Salons et journaux: Souvenirs des milieux littéraires, politiques, artistiques et médicaux de 1880 à 1908* (Paris: Nouvelle Librairie Nationale, 1917), 219-220. Usually brother Armand does not come off well in biographical sketches of the Reclus family because he was farthest to the right politically, but Elie Faure, at least, thought that Armand had a formidable intellect. Letter from Elie Faure to Paul Deschamps, 18 January 1921, *Oeuvres complètes*, vol. 3, *Oeuvres diverses* (Paris: Jean-Jacques Pauvert Editeur, 1964), 1024-1025.

2. Paul's book, completed in 1939, was published as *Les Frères Elie et Elisée Reclus; ou, du protestantisme à l'anarchisme* (Paris: Les Amis d'Elisée Reclus, 1964). "Les Amis" were sons Michel and Jacques, plus a number of old anarchists, including some Spanish exiles. The work contains a biographical sketch of Paul on p. 203. See also his autobiographical essay, "Synthèse d'un individu," *Plus Loin*, 14th year, no. 156 (April 1938), 1-5. Very revealing of Paul's character are the letters which he wrote to Joseph Ishill between 1923 and 1935 (Joseph Ishill Collection, Houghton Library, Harvard University).

For the life and work of Joseph Ishill, see the following articles in *The American Book Collector*: Marian Courtney Brown, "The Oriole Press. Joseph Ishill: Solitary Rebel," vol. 13, no. 1 (September 1962), 8-16; William White, "Joseph Ishill (1888-1966)," vol. 16, no. 7 (March 1966), 20; and Mrs. Simon Mendelsohn (Ishill's daughter), "A Complete Checklist of the Publications of Joseph Ishill and His Oriole Press," vol. 25, no. 1 (September-October 1974), 14-25, no. 2 (November-December 1974), 20-31, and no. 3 (January-February 1975), 16-23.

3. Biographical data on Elisée Reclus's descendants were supplied by his grandson, Jacques Régnier of Marseilles, and great-granddaughters Louise Rapacka of Paris and Jeannie Geddes of Edinburgh.

Like his grandfather, Louis Cuisinier enjoyed hiking in mountainous areas, but he literally went farther and became a true mountain-climber. In 1925 he claimed to be the first Frenchman to climb Mount Kilimanjaro. See his paper, "Les Neiges de l'Equateur," *L'Illustration*, vol. 168, no. 4354 (14 August 1926), 158-159.

For the Shalit case, in which Anne Geddes Shalit and her husband Binyamin successfully petitioned the Israel High Court of Justice to have their children registered as Jews, see Amnon Rubinstein, "Who's a Jew, and Other Woes," *Encounter*, vol. 36, no. 3 (March 1971), 84-93.

4. Edmund Wilson, *To the Finland Station: A Study in the Writing and Acting of History* (New York: Farrar, Straus and Giroux, 1972, reprint of 1940 edition with new introduction). On the matter of the influence of Reclus and other anarchist writers on Stalin, Mao, and the Flores Magon brothers, see Y. G. Saushkin, "The Geographical

Environment of Human Society," *Soviet Geography: Review and Translation*, vol. 4, no. 10 (December 1963), 13; Ian M. Matley, "The Marxist Approach to the Geographical Environment," *Annals of the Association of American Geographers*, vol. 56, no. 1 (March 1966), 108; Robert A. Scalapino and George T. Yu, *The Chinese Anarchist Movement* (Berkeley: University of California, Institute of International Studies, Center for Chinese Studies, 1961), 1, 6; Marcelo Segall. "En Amérique latine. Développement du mouvement ouvrier et proscription," *International Review of Social History*, vol. 17, parts 1-2 (1972), 330. For Blasco Ibáñez's translations and publications of Reclus's works, see Paul C. Smith, *Vicente Blasco Ibáñez: An Annotated Bibliography* (London: Grant and Cutler, 1976), 41.

5. Of this group, Nieuwenhuis may be the least well known, except in the Netherlands, where his memory has been kept alive by Dutch leftists, including the students who tried to rename the University of Amsterdam for him in 1969. Ferdinand Domela Nieuwenhuis (1846-1919) was strikingly similar to Elisée Reclus, even in his personal life. Sons of Protestant ministers, both disavowed formal religion but continued as missionaries of a new faith. They were veritable messiahs of anarchism, which is an anti-messianic and anti-authoritarian ideology. See Rudolf de Jong, "Ferdinand Domela Nieuwenhuis: Anarchist and Messiah," *Delta*, vol. 13, no. 4 (Winter 1970-1971), 65-78.

6. Will and Ariel Durant, *A Dual Autobiography* (New York: Simon and Schuster, 1977), 27, 43, 69; Italo Calvino, "Un pomeriggio, Adamo," in *I Racconti* (Turin: Einaudi, 1958), 24, and *Adam, One Afternoon and Other Stories*, translated by Archibald Colquhoun and Peggy Wright (London: Collins, 1957), 15. I am indebted to Consuelo Dutschke for bringing the Italian passage to my attention.

7. R. J. Harrison Church, "The French School of Geography," Chapter 3 of *Geography in the Twentieth Century*, ed. by Griffith Taylor (New York: The Philosophical Library, 1951), 71-72; André Meynier, *Histoire de la pensée géographique en France (1872-1969)* (Paris: Presses Universitaires de France, 1969), 11.

8. Professor Aimé Perpillou, in letter, 9 October 1971.

9. Charles Fisher in *Geographers Abroad: Essays on the Problems and Prospects of Research in Foreign Areas*, ed. by Marvin Mikesell (University of Chicago, Department of Geography, Research Paper No. 152, 1973), 202. Reclus's treatment of India in the NGU has been singled out by several modern scholars for its enduring quality. See David Sopher and Charles Fisher in *Geographers Abroad*, 116-117, 202; and O. H. K. Spate, *India and Pakistan: A General and Regional Geography*, 2nd ed. (London: Methuen & Co., Ltd., 1957), vii. Spate has said, " . . . for grasp of fundamental relationships, for really masterly presentation of broad essentials, I know of no geographer superior to Reclus."

Olinto Marinelli's views were published in his obituary of Reclus: "Eliseo Reclus," *Il Marzocco*, vol. 10, no. 28 (9 July 1905), unpaged (pp. 2-3).

10. Luigi Filippo de Magistris, "Perchè veneriamo Eliseo Reclus," *La Geografia*, vol. 4, no. 10 (1916), 432; Kenneth Rexroth, "Revolution Now!," *San Francisco*, vol. 11, no. 12 (December 1969), 19-20; Rexroth, *Communalism: From Its Origins to the Twentieth Century* (New York: The Seabury Press, 1974), xiii; George Woodcock, "Anarchism and Ecology," *The Ecologist*, vol. 4, no. 3 (March-April 1974), 84-88; D. R. Stoddart, "Kropotkin, Reclus, and 'Relevant' Geography," *Area*, vol. 7, no. 3 (1975), 188-190.

11. Meynier, *op. cit.*, 13.

NOTES

12. Béatrice Giblin, "Elisée Reclus. Pour une géographie," Unpublished thesis (3rd cycle doctorate), University of Paris-Vincennes, 1971, 112, 219ff.; Giblin, "Elisée Reclus: géographie, anarchisme," *Hérodote*, no. 2 (April-June 1976), 30ff.; Paul B. Sears, "Ecology — A Subversive Science," *BioScience*, vol. 14, no. 7 (July 1964), 11; cf. Paul Shepard and Daniel McKinley, *The Subversive Science: Essays toward an Ecology of Man* (Boston: Houghton, Mifflin Company, 1969).

Bibliography

Unpublished Materials

Archives

Belgium

Brussels
    Académie Royale des Sciences, des Lettres et des Beaux-Arts de Belgique
        A miscellany of materials relating to Elisée Reclus
    Université Libre de Bruxelles, Service des Archives
        Archives of the Université Nouvelle

France

Paris
    Archives Librairie Hachette
        Dossier on Elisée Reclus
    Archives Nationales
        Dossier on Elisée Reclus (BB^{24}/732)
    Archives Préfecture de Police
        Two folders of materials on Elisée Reclus (B a/1.237 and E a/103 22)
        Box E A/129 contains photographs of anarchists, as well as

two copies each of two issues (August and September 1894) of *Album photographique des individus qui doivent être l'objet d'une surveillance spéciale aux frontières*. The *Albums* give names of Elisée, Elie, and Paul (Elie's son), a description of Paul, and photographs of Elisée and Paul.

Bibliothèque Nationale, Département des manuscrits

N.A.F. (Nouvelles Acquisitions Françaises) 22909-22919 (11 volumes). Correspondance et papiers d'Elisée Reclus.

N. A. F. 24282. Autographes Félix et Paul Nadar, XXIII, Raban-Reclus.

Fonds Rapacka

Collection of private papers in the possession of Mme. Louise Rapacka, a great-granddaughter of Elisée Reclus.

Institut français d'histoire sociale

Fonds Elisée Reclus (14/AS/232)

Saint-Denis

Bibliothèque Municipale de Saint-Denis

R^{c}MS 68. "Rapport du Capitaine Commandant par interim le 119e Bon de la Garde Nationale de la Seine sur la journée sur la journée [sic] du 4 avril. Affaire de Châtillon."

Sainte-Foy-la-Grande

Hôtel de Ville

Birth registers. Elisée Reclus' birth is recorded in a volume labeled "Naissances. 1833 [sic] à 1842" (apparently misdated when bound), section "An 1830," page 6.

Fonds Corriger

Jean Corriger, "Les 3 frères Faure," unpublished manuscript (dated 28 August 1966) in the possession of Mme. Corriger

The Netherlands

Amsterdam

Internationaal Instituut voor Sociale Geschiedenis

BIBLIOGRAPHY

Nettlau Archive (Reclus Dossiers)
Archief Elisée Reclus

Switzerland

Geneva
Bibliothèque publique et universitaire de Genève, Département des
manuscrits
Numerous Reclus items scattered through several collections
(*e. g.*, Baud-Bovy and Chaponnière). Professor Vincent
Berdoulay made notes and photocopies for me.

Neuchâtel
Société neuchâteloise de géographie, Archives (located in the attic of
the Institut de géologie of the Université de Neuchâtel)
Numerous Reclus items, including the manuscript (incomplete) of the *Nouvelle géographie universelle*. Also correspondence between Elisée Reclus and Charles Knapp,
archiviste-bibliothécaire of the Société neuchâteloise de
géographie. Professor Berdoulay made notes and photocopies for me.

United Kingdom

Edinburgh
National Library of Scotland, Department of Manuscripts
Patrick Geddes Papers

London
Royal Geographical Society
Eight letters from Elisée Reclus

United States

Burlington, Vermont
University of Vermont, Bailey Library, Special Collections
George Perkins Marsh Collection

Cambridge, Massachusetts
Harvard University, Houghton Library
Joseph Ishill Collection
Radcliffe College, The Arthur and Elizabeth Schlesinger Library on the
History of Women in America
Mary Putnam Jacobi Papers

Denton, Texas
Engerrand Papers (held by Mrs. Anita Engerrand Gafford)
Papers of George Engerrand. Materials relating to Elisée Reclus, including Engerrand's notes for a projected biography of Reclus and also correspondence from Reclus to Engerrand and Jeanne Richard (later Mrs. Engerrand), 1898-1905.

New Orleans, Louisiana
Louisiana State Museum Library
George C. H. Kernion, "Captain Michel Jean Fortier, A Speech," Unpublished typescript dated 16 April 1931.
Marvin Rogers, "Captain Michael [sic] Jean Fortier, 1843-1883," Unpublished typescript dated 25 September 1937.
Tulane University, Howard-Tilton Library, Special Collections Division
R. P. Martin, Jr., "The Plantation Mansion and Estate of Valcour Aime . . .," Unpublished typescript dated May 1968.
United States Census, 7th Census (1850), Louisiana — Film 208 (St. James Parish)
Work Projects Administration, Survey of Federal Archives in Louisiana, "Passenger Lists Taken from Manifests of the Customs Service, Port of New Orleans, 1850-1861" (Book IV), Unpublished typescript dated 1941.
Tulane University, School of Medicine, Rudolph Matas Medical Library.
"Registre du Comité Médical de la N^{ell} Orléans" (1816-1854).
"St. James Parish Medical Society and the History of Medical Organization in the Parish," Unpublished typescript in folder labeled "St. James Parish" in Rudolph Matas Papers

Washington, D. C.
  National Archives
    Record Group 84, Microcopy No. T-1, Despatches from U. S.
      Consuls in Paris, 1790-1806, Roll 13, vol. 13 (1 July
      1861-25 January 1864), letters from John Bigelow to
      William Seward, 29 May and 3 July 1863

Unpublished Theses

VINCENT RAYMOND HENRI BERDOULAY, "The Emergence of the
  French School of Geography (1870-1914)," Ph.D. dissertation in
  Geography, University of California, Berkeley, 1974.
BÉATRICE GIBLIN, "Elisée Reclus. Pour une géographie," Thesis (Doc-
  torat de troisième cycle), University of Paris-Vincennes, 1971.
BENJAMIN HARRISON, "Gabriel Monod and the Professionalization of
  History in France, 1844-1912," Ph.D. dissertation in History, Uni-
  versity of Wisconsin, 1972.
JAMES R. KROGZEMIS, "A Historical Geography of the Santa Marta
  Area, Colombia," Ph.D. dissertation in Geography, University of
  California, Berkeley, 1967 (A number of copies were made for the
  Office of Naval Research distribution list).
JOHN P. REILLY, "The Early Social Thought of Patrick Geddes," Ph.D.
  dissertation in History, Columbia University, 1971.

Biographical Works
On Elisée Reclus

No attempt is made here to cite brief encyclopedia articles or incidental references to Reclus in recent histories of geography. Such works are often derivative and inaccurate and offer no real insights into Reclus's life and work. An important exception would be the valuable discussion of Reclus and his works in *Histoire de la pensée géographique en France* (1872-1969) by André Meynier (Paris: Presses Universitaires de France, 1969).

ANONYMOUS. "M. Elisée Reclus and the *Géographie Universelle.*" *Scottish Geographical Magazine*, vol. 11, no. 5 (May 1895), 248-251.

GABRIEL ASTRUC. "La Famille Reclus." *L'Illustration*, vol. 102, no. 2652 (23 December 1883), 569.

ALDO BLESSICH. "L'Opera di Eliseo Reclus." *Bollettino della Società Geografica Italiana*, series 4, vol. 6, no. 8 (August 1905), 579-600.

GIUSEPPE CARACI. "Prefazione." Pp. 5-30 of Caraci's translation of Reclus's *Histoire d'un ruisseau* (*Storia di un ruscello*). 3rd edition. Florence: "La Nuova Italia" Editrice, 1933.

A. C. A. COMPÈRE-MOREL. *Grand dictionnaire socialiste*. Paris: Publications sociales, 1924. Reclus on p. 711.

HEM DAY (pseudonym of Marcel Dieu) et al. "Elisée Reclus (1830-1905), savant et anarchiste." *Les Cahiers Pensée et Action* (Paris and Brussels), no. 5 (April-June 1956). Parts issued separately, *Essai de bibliographie de Elisée Reclus* and *Elisée Reclus en Belgique: Sa vie, son activité, 1894-1905*.

CH. D. [Charles Delfosse] "Elisée Reclus, géographe." *Les Hommes du jour*, 1st series, no. 32a, n. d. (1895-1896).

G. S. DUNBAR. "Elisée Reclus." *Dictionary of Scientific Biography*, vol. 11 (New York: Scribner's, 1975), 337-338.

JACQUES DUZER. *Les Idées anarchistes d'Elisée Reclus*. Dijon, Faculté de droit et des sciences économiques de Dijon, Diplome d'études supérieures d'histoire du droit et des faits sociaux, Mémoire, 1969 (mimeographed). "Biographie" on pp. 5-30.

BÉATRICE GIBLIN. "Elisée Reclus: géographie, anarchisme." *Hérodote*, no. 2 (April-June 1976), 30-49.

Y.–M. GOBLET. "Un grand centenaire géographique: Elisée Reclus." *La Géographie*, vol. 55, nos. 1-2 (January-February 1931), 72-75.

JOSEPH ISHILL, compiler, editor, and printer. *Elisée and Elie Reclus: In Memoriam*. Berkeley Heights, New Jersey: The Oriole Press, 1927.

DAGMAR RENSHAW LEBRETON. "Un Anarchiste sur une plantation louisianaise en 1855." *Comptes Rendus de l'Athénée Louisianais* (New Orleans) (March 1954), 27-32.

LUIGI FILIPPO DE MAGISTRIS. "Perchè veneriamo Eliseo Reclus." *La Geografia*, vol. 4, no. 10 (1916), 431-432.

JEAN MAITRON, ed. *Dictionnaire biographique du mouvement ouvrier français*. Volume 8, *Deuxième partie: 1864-1871, La Première Internationale et la Commune, Mor à Rob*. Paris: Les Editions Ouvrières, 1970. Reclus on pp. 299-301.

R. MARZAC. "Les Reclus." *Le Figaro* (Paris), 40th year, 3rd series, no. 211 (30 July 1894), 1-2.

BIBLIOGRAPHY

CLARA MESNIL. "Souvenirs sur Elisée Reclus." *Maintenant* (Paris), no. 2 (1946), 252-256.

MARVIN W. MIKESELL. "Observations on the Writings of Elisée Reclus." *Geography*, vol. 44, part 4 (November 1959), 221-226.

MAX NETTLAU. *Elisée Reclus, Anarchist und Gelehrter (1830-1905)*. (Beiträge zur Geschichte des Sozialismus, Syndikalismus, Anarchismus, Vol. 4) Berlin: Verlag "Der Syndikalist," Fritz Kater, 1928. Also published in an enlarged Spanish edition, *Eliseo Reclus, la vida de un sabio justo y rebelde*, translated by V. Orobón Fernández, 2 vols. ("Biblioteca de La Revista Blanca") (Barcelona: Publicaciones de "La Revista Blanca," n. d. [1929]).

MAX NETTLAU. "Jacques Elisée Reclus." *Encyclopaedia of the Social Sciences*, vol. 13 (London: Macmillan and Co., Ltd., 1934), 164-165.

MAURICE PEYROT. "Elisée Reclus." *La Nouvelle Revue*, vol. 50, no. 1 (1 January 1888), 170-185.

Y. R. (identity unknown). "Un cinquantenaire: Elisée Reclus." *Le Monde* (Paris), 12th year, no. 3248 (5 July 1955), 6.

PAUL RECLUS. *Les Frères Elie et Elisée Reclus: ou, du protestantisme à l'anarchisme*. Paris: Les Amis d'Elisée Reclus, 1964.

H. ROORDA VAN EYSINGA. "Elisée Reclus propagandiste." *La Société nouvelle, revue internationale* (Mons and Paris), 2nd series, 13th year, no. 2 (August 1907), 186-199.

HAN RYNER (pseudonym of Henry Ner). "Elisée Reclus (1830-1905)." (La "Bonne Collection," no. 77) *La Brochure mensuelle* (Paris), no. 61 (January 1928).

Le Semeur (Caen), Numéro spécial (8 February 1928). 8 pages. Issue devoted to articles about Elisée Reclus.

JEAN STEENS. "Profils socialistes." *Le Correspondant*, new series, vol. 175, no. 6 (25 June 1903), 1165-1190. Reclus on pp. 1176-1178.

FRANÇOIS STOCKMANS. [Article on Elisée Reclus] *Biographie Nationale*, Vol. 34, Supplement: Vol. 6 (Fascicule 2), columns 671-690. Brussels: L'Académie Royale des Sciences, des Lettres et des Beaux-Arts de Belgique, 1968.

G. S. TIKHOMIROV. "Elize Rekliu-vydaiushchiisia frantsuzskii geograf." *Istoriia geologo-geograficheskikh nauk*, no. 2 (Trudy Instituta Istorii Estestvoznaniia i Tekhniki, vol. 37, pp. 38-51). Moscow, 1961.

HELEN ZIMMERN. "Elisée Reclus and His Opinions." *The Popular Science Monthly*, vol. 44 (1894), 402-408.

Obituaries
of Elisée Reclus

This is by no means a complete list of Reclus obituaries, but I have selected several that I think are of special significance. Particularly important are those obituaries written by Reclus's friends and associates. I am not listing newspaper articles because they are usually anonymous, mostly derivative, frequently erroneous, and generally so brief that they do not provide the insights that occur in journal (revue) articles.

G. A. [Gunnar Andersson] "Elisée Reclus." *Ymer*, vol. 24, no. 3 (1905), 325-329.

ANONYMOUS. "Elisée Reclus." *L'Illustration*, vol. 126. no. 3254 (8 July 1905), 32.

ANONYMOUS. "Elisée Reclus—Geographer, Philosopher, Anarchist." *Current Literature*, vol. 39, no. 5 (November 1905), 563-565.

ANONYMOUS. "Elisée Reclus, 1830-1905." Reprint from No. 9-10 (1905) of *Bulletin de la Société belge d'astronomie*.

ANONYMOUS. "Mort d'Elisée Reclus." *A Travers le monde*, new series, vol. 11, no. 29 (22 July 1905), 230.

ANONYMOUS. "Professor Elisée Reclus." *Bulletin of the American Geographical Society*, vol. 37, no. 8 (1905), 496-497.

ANONYMOUS. "Science Gossip." *The Athenaeum*, no. 4054 (8 July 1905), 55.

ANONYMOUS. [No title] *Bulletin de la Société de géographie et d'études coloniales de Marseilles*, vol. 29, no. 3 (3me Trimestre 1905), 322-323.

M. B. [Marcellin Boule] "Mort d'Elisée Reclus." *L'Anthropologie*, vol. 16 (1905), 596.

E. C. [E. Cammaerts] "Elisée Reclus." *Bulletin de la Société royale belge de géographie*, vol. 29 (1905), 210-215.

L. G. [Lucien Gallois] "Elisée Reclus." *Annales de géographie*, vol. 14, no. 76 (15 July 1905), 373-374.

PATRICK GEDDES. "A Great Geographer: Elisée Reclus, 1830-1905." *Scottish Geographical Magazine*, vol. 21, no. 9 (September 1905), 490-496; no. 10 (October 1905), 548-555.

PAUL GHIO. *En Souvenir d'Elisée Reclus.* (Causerie faite au Chateau du Peuple le 20 aôut 1905). Paris: L'Emancipatrice (imprimerie communiste), 1905.

PAUL GIRARDIN AND JEAN BRUNHES. "Elisée Reclus' Leben and Wirken (1830-1905)." *Geographische Zeitschrift*, vol. 12, no. 2 (February 1906), 65-79. Also published in French as "La Vie et l'oeuvre d'Elisée

Reclus (1830-1905)," *Revue de Fribourg,* vol. 37, no. 4 (April 1906), 274-287; no. 5 (May 1906), 355-365.

GUILLAUME DE GREEF. *Eloges d'Elisée Reclus et de Kellès-Krauz.* (Discours prononcé par Monsieur le recteur Guillaume De Greef, Séance de rentrée du 3 novembre 1905, Université Nouvelle de Bruxelles, Institut des hautes études) Ghent: Société Coopérative "Volksdrukkerij," 1906.

HERMANN HAACK. "Die Toten des Jahres 1905." Pp. 191-257 of *Geographen-Kalender: Vierter Jahrgang 1906/1907.* Gotha: Justus Perthes, 1906. Reclus on pp. 231-232.

RICHARD HEATH. "Elisée Reclus." *The Humane Review,* vol. 6 (October 1905), 129-142.

C. M. KAN. "Elisée Reclus in het kader der geografen van zijnen tijd (1830-1905)." *Tijdschrift van het Koninklijk Nederlandsch Aardrijkskundig Genootschap,* second series, vol. 22, part 2 (1905), 1033-1051.

C. KNAPP. "Elisée Reclus." *Bulletin de la Société neuchâteloise de géographie,* vol. 16 (1905), 310-316.

P. KROPOTKIN. "Elisée Reclus." *Geographical Journal,* vol. 26, no. 3 (September 1905), 337-343.

OLINTO MARINELLI. "Eliseo Reclus." *Il Marzocco,* vol. 10, no. 28 (9 July 1905), unpaged [pp. 2-3].

JACQUES MESNIL. *Eliseo Reclus.* Mantova: Stab. Tip. Baraldi & Fleischmann, 1905. Reprinted from *Il Pensiero* of Rome.

G. RICCHIERI. "Eliseo Reclus." *Rivista Geografica Italiana,* vol. 13, no. 2-3 (February-March 1906), 113-125.

G. RONCAGLI. "Eliseo Reclus." *Bollettino della Società Geografica Italiana,* series 4, vol. 6, no. 8 (August 1905), 573-574.

F. SCHRADER. "Elisée Reclus." *La Géographie,* vol. 12, no. 2 (15 August 1905), 81-86.

W. WOLKENHAUER. "Elisée Reclus." *Deutsche Rundschau für Geographie und Statistik,* vol. 28, no. 1 (October 1905), 36-40.

A BIBLIOGRAPHY OF THE WRITINGS
OF ELISEE RECLUS

This bibliography is not perfectly complete, but it consists of works actually seen by me. I have probably included everything of geographical significance. The chronological ordering of Reclus's writings gives a sense of the evolution of his social philosophy.

The interested reader can find additional items, including various editions and translations, in the following bibliographies:

Bibliographie de la France. Journal général de l'imprimerie et de la librairie. Paris: Pillet, etc., 1811-

HEM DAY (pseudonym of Marcel Dieu). *Essai de bibliographie de Elisée Reclus.* Paris and Brussels: Pensée et Action, 1956.

[Louise Dumesnil] "Elisée Reclus, géographie et sociologie." Unpaged typescript with manuscript additions. N. p., n.d. Copy in the department of Printed Books, Bibliothèque Nationale, Paris.

Ente per la Storia del Socialismo e del Movimento Operaio Italiano (E.S.M.O.I.). *Bibliografia del Socialismo e del Movimento Operaio Italiano.* 2 vols. in 7 parts. Rome and Turin: Edizioni E.S.M.O.I., 1956-1968. For Reclus, see vol. 2, part 3 (1966), 295-297.

International Institute of Social History, Amsterdam (Internationaal Instituut voor Sociale Geschiedenis). *Alphabetical Catalogue of the Books and Pamphlets (Alfabetische Catalogus van de Boeken en Brochures).* 12 vols. Boston: G. K. Hall & Co., 1970. For Reclus, see vol. 9, 633-641.

OTTO LORENZ, etc., eds. *Catalogue général de la librairie française.* 34 vols. Paris: Chez O. Lorenz, etc., 1867-1945.

JEAN MAITRON. *Histoire du mouvement anarchiste en France (1880-1914).* Paris: Société universitaire d'éditions et de librairie, 1951. Pp. 539-716, "Bibliographie du mouvement anarchiste en France, 1880-1914." Reclus items on pp. 684-691.

Ministère de l'éducation nationale. *Catalogue général des livres imprimés de la Bibliothèque Nationale.* Vol. 147. Paris: Imprimerie nationale, 1938. Reclus items on columns 650-676.

MAX NETTLAU. *Bibliographie de l'anarchie.* (Bibliothèque des "Temps Nouveaux," Année 1897, no. 8) Brussels: Bibliothèque des "Temps Nouveaux"; and Paris; P.–V. Stock, 1897. Preface by Elisée Reclus. Reclus items on pp. 67-71, 237.

JOSEF STAMMHAMMER. *Bibliographie des Socialismus und Communismus.* 3 vols. Jena: G. Fischer, 1893-1909 (Reprinted Aalen: Otter Zeller, 1963). For Reclus, see vol. 1, 197; vol. 2, 271; and vol. 3, 278.

HUGO P. THIEME. *Bibliographie de la littérature française de 1800 à 1930.* 2 vols. Paris: Librairie E. Droz, 1933. Reclus items in vol. 2, 575-576.

1857   "Lettres d'un voyageur." *L'Union* (New Orleans newspaper), vol. 1, no. 7 (7 February 1857), [4]; no. 8 (8 February 1857), [4]; no. 9

(9 February 1857), [4]; no. 16 (16 February 1857), [4]; no. 17 (17 February 1857), [4].

"Nouvelle-Grenade." *L'Union,* vol. 1, no. 175 (26 July 1857), 2-3; no. 178 (29 July 1857), 2-3; no. 179 (30 July 1857), 2-3; no. 180 (31 July 1857), 2; no. 181 (1 August 1857), 2-3; no. 183 (3 August 1857), 2-3; no. 185 (5 August 1857), 2-3; no. 186 (6 August 1857), 2; no. 188 (8 August 1857), 2-3; no. 190 (10 August 1857), 2-3; no. 191 (11 August 1857), 3.

1858   "Considérations sur quelques faits de géologie et d'ethnologie." *Revue philosophique et religieuse,* vol. 9, no. 34 (1 January 1858), 218-227. [Review article based on *Histoire du sol de l'Europe* by J. C. Houzeau (Brussels: Librairie internationale, 1857)]

1859   "Quelques mots sur la Nouvelle-Grenade." *Bulletin de la Société de géographie* (Paris), 4th series, vol. 17, nos. 97-98 (January-February 1859), 111-141.

"Le Mississipi, Etudes et souvenirs.—I.—Le Cours supérieur du fleuve." *Revue des Deux Mondes,* vol. 22 (15 July 1859), 257-296.

"Le Mississipi, Etudes et souvenirs.—II.—Le Delta et la Nouvelle-Orléans." *Revue des Deux Mondes,* vol. 22 (1 August 1859), 608-647.

"Etude sur les fleuves." *Bulletin de la Société de géographie,* 4th series, vol. 18, no. 104 (August 1859), 69-104.

"La Nouvelle-Grenade, paysages de la nature tropicale.—I.—Les Côtes néo-grenadines." *Revue des Deux Mondes,* vol. 24 (1 December 1859), 624-661.

Translation and introduction of Carl Ritter, "De la configuration des continents sur la surface du globe, et de leurs fonctions dans l'histoire." *Revue germanique,* vol. 8, no. 11 (1859), 241-267.

1860   *Guide du voyageur à Londres et aux environs.* (Collection des Guides-Joanne) Paris: Librairie de L. Hachette et Cie, n. d. Abridged version, 1862, published as *Londres illustré, guide spécial pour l'exposition de 1862.*

"La Nouvelle-Grenade, paysages de la nature tropicale.—II.—Sainte-Marthe et la Horqueta." *Revue des Deux Mondes*, vol. 25 (1 February 1860), 609-635.

"La Nouvelle-Grenade, paysages de la nature tropicale.—III.—Rio-Hacha, les Indiens Goajires et la Sierra-Negra." *Revue des Deux Mondes*, vol. 26 (15 March 1860), 419-452.

"La Nouvelle-Grenade, paysages de la nature tropicale.—IV.—Les Aruaques et la Sierra-Nevada." *Revue des Deux Mondes*, vol. 27 (1 May 1860), 50-83.

"Fragment d'un voyage à la Nouvelle-Orléans." *Le Tour du Monde*, vol. 1 (1860), 177-192.

"Excursion dans le Dauphiné." *Le Tour du Monde*, vol. 2, no. 52 (1860), 402-416.

"Voyage de M. Du Chaillu dans l'Afrique occidentale." *Bulletin de la Société de géographie*, 4th series, vol. 20 (October 1860), 271-275. [A report of a recent paper by Du Chaillu at the American Geographical Society]

"De l'Esclavage aux Etats-Unis.—I.—Le Code noir et les esclaves." *Revue des Deux Mondes*, vol. 30 (15 December 1860), 868-901. [Review article]

1861   "De l'Esclavage aux Etats-Unis.—II.—Les Planteurs et les abolitionistes." *Revue des Deux Mondes*, vol. 31 (1 January 1861), 118-154. [Review Article]

"Le Mormonisme et les Etats-Unis." *Revue des Deux Mondes*, vol. 32 (15 April 1861), 881-914. [Review Article].

"Paysages du Taurus Cilicien." *Revue germanique*, vol. 15, no. 1 (1 May 1861), 43-60. [Review article]

"La Méditerranée caspienne et le Canal des steppes." *Revue des Deux Mondes*, vol. 34 (1 August 1861), 592-623. [Review article]

Voyage à la Sierra-Nevada de Sainte-Marthe. Paysages de la nature tropicale. ("Bibliothèque des chemins de fer") Paris: Librairie de L. Hachette et Cie.

1862   "Le Coton et la crise américaine." *Revue des Deux Mondes*, vol. 37 (1 January 1862), 176-208.

"Ensayo sobre las revoluciones políticas y la condicion social de las repúblicas colombianas, por José M. Samper." *Bulletin de la Société de géographie*, series 5, vol. 3 (February 1862), 96-112. [Review article]

"Les Cités lacustres de la Suisse.—Un peuple retrouvé." *Revue des Deux Mondes*, vol. 37 (15 February 1862), 883-902. [Review article]

"Atlas sphéroïdal et universel de géographie, par F. A. Garnier." *Bulletin de la Société de géographie*, 5th series, vol. 3 (March 1862), 177-182. [Review article]

"Introduction." Pp. xvii-xlvi of Adolphe Joanne, *Itinéraire général de la France*, Volume 3, *Les Pyrénées et le réseau des chemins de fer du Midi et des Pyrénées*, 2nd ed. ("Collection des Guides-Joanne") Paris: Librairie de L. Hachette et Cie.

"Le Brésil et la colonisation.—I.—Le Bassin des Amazones et les Indiens." *Revue des Deux Mondes*, vol. 39 (15 June 1862), 930-959. [Review article]

"Le Brésil et la colonisation.—II.—Les Provinces du littoral, les noirs et les colonies allemandes." *Revue des Deux Mondes*, vol. 40 (15 July 1862), 375-414. [Review article]

"Les Livres sur la crise américaine." *Revue des Deux Mondes*, vol. 42, no. 2 (15 November 1862), 505-512. [Review article]

"Le Littoral de la France.—L'Embouchure de la Gironde et la péninsule de Grave." *Revue des Deux Mondes*, vol. 42 (15 December 1862), 901-936.

1863   "Un Prisonnier de guerre au Mexique." *Revue des Deux Mondes*, vol. 43 (1 February 1863), 765-768. [Review]

"Report on the physics and hydraulics of the Mississipi [sic] *River* . . . by Captain A. A. Humphreys and Lieutenant H. L. Abbot . . ." *Bulletin de la Société de géographie*, 5th series, vol. 5 (February 1863), 126-161. [Review article]

"Un Voyage dans la Tunisie." *Revue des Deux Mondes*, vol. 44 (1 March 1863), 249-252. [Review]

"Les Noirs américains depuis la guerre civile des Etats-Unis.—I.—Les Partisans du Kansas et les noirs libres de Beaufort." *Revue des Deux Mondes*, vol. 44 (15 March 1863), 364-394. [Review article]

"Les Noirs américains depuis la guerre.—II.—Les Plantations de la Louisiane.—Les Régimens africains.—Les Décrets d'émancipation." *Revue des Deux Mondes*, vol. 44 (1 April 1863), 691-722.

"Le Littoral de la France.—II.—Les Landes du Médoc et les dunes de la côte." *Revue des Deux Mondes*, vol. 46 (1 August 1863), 673-702.

"Recherches sur les ouragans." *Revue des Deux Mondes*, vol. 46 (15 August 1863), 1017-1019.[Review]

"Le Littoral de la France.—III.—Les Plages et le Bassin d'Arcachon." *Revue des Deux Mondes*, vol. 48 (15 November 1863), 460-491.

1864   "La Poésie et les poètes dans l'Amérique espagnole." *Revue des Deux Mondes*, vol. 49 (15 February 1864), 902-929. [Review article]

"Un Ecrit américain sur l'esclavage." *Revue des Deux Mondes*, vol. 50 (15 March 1864), 507-510. [Review]

"La Commission sanitaire de la guerre aux Etats-Unis." *Revue des Deux Mondes*, vol. 51 (1 May 1864), 155-172. [Review article]

"Le Littoral de la France.—IV.—Les Landes de Born et du Marensin." *Revue des Deux Mondes*, vol. 53 (1 September 1864), 191-217.

"Deux années de la grande lutte américaine." *Revue des Deux Mondes*, vol. 53 (1 October 1864), 555-624.

"L'Homme et la Nature.—De l'Action humaine sur la géographie physique." *Revue des Deux Mondes*, vol. 54 (15 December 1864), 762-771. [Review of *Man and Nature* by G. P. Marsh]

BIBLIOGRAPHY

Les Villes d'hiver de la Méditerranée et les Alpes maritimes. ("Collection des Guides-Joanne") Paris: Hachette.

"Introduction." Pp. xvii-clix of Adolphe Joanne, *Dictionnaire des communes de la France,* vol. 1, Paris: Hachette.

1865   "Les Oscillations du sol terrestre." *Revue des Deux Mondes,* vol. 55 (1 January 1865), 57-84. [Review article]

"Histoire du peuple américain, par Auguste Carlier." *Bulletin de la Société de géographie,* 5th series, vol. 19 (February 1865), 143-164. [Review]

"La Guerre de l'Uruguay et les Républiques de la Plata." *Revue des Deux Mondes,* vol. 55 (15 February 1865), 967-997. [Review article]

Review of *Annuaire scientifique* (vol. 4, 1865), published by Dehérain. *Nouvelles annales des voyages, de la géographie, de l'histoire et de l'archéologie,* vol. 185 (6th series, 11th year, 1865, vol. 1, February 1865), 232-233. [Review]

"Les Fleuves." *Nouvelles annales des voyages . . .,* vol. 185 (6th series, 11th year, 1865, vol. 1, March 1865), 257-299; vol. 186 (1865, vol. 2, April 1865), 24-63.

"Etudes sur les dunes." *Bulletin de la Société de géographie,* 5th series, vol. 9 (March 1865), 193-221.

"Le Mont Etna et l'éruption de 1865." *Revue des Deux Mondes,* vol. 58 (1 July 1865), 110-138.

1866   Review of M. de Tchihatchef, *Le Bosphore et Constantinople avec perspective des pays limitrophes* (Paris: Morgand, 1865). *Revue des Deux Mondes,* vol. 61 (1 January 1866), 262-267.

"Les Estuaires et les deltas. Etude de géographie physique." *Annales des voyages . . .,* vol. 190 (Année 1866, vol. 2, April 1866), 5-55.

"Du Sentiment de la nature dans les sociétés modernes." *Revue des Deux Mondes,* vol. 63 (15 May 1866), 352-381.

"Atlas de la Colombie, publié par ordre du gouvernement colombien." *Bulletin de la Société de géographie*, 5th series, vol. 12 (August 1866), 140-146. [Review].

"Les Républiques de l'Amérique du Sud, leurs guerres et leur projet de fédération." *Revue des Deux Mondes*, vol. 65 (15 October 1866), 953-980. [Review article].

"La Sicile et l'éruption de l'Etna en 1865." *Le Tour de Monde*, vol. 13 (1866), 353-416.

1867   "Les Forces souterraines." *Revue des Deux Mondes*, vol. 67 (1 January 1867), 218-230. [Review]

"Les Plages et les fiords." *Revue des Deux Mondes*, vol. 68 (1 March 1867), 265-272.

"Les Basques, un peuple qui s'en va." *Revue des Deux Mondes*, vol. 68 (15 March 1867), 313-340. [Review article]

"John Brown." *La Coopération* (14 July 1867), reprint in Bibliothèque Nationale.

"L'Océan, étude de physique maritime." *Revue des Deux Mondes*, vol. 70 (15 August 1867), 963-993. [Review article]

"La Guerre du Paraguay." *Revue des Deux Mondes*, vol. 72 (15 December 1867), 934-965.

1868   "Les Républiques de l'isthme américain." *Revue des Deux Mondes*, vol. 74 (15 March 1868), 479-498. [Review]

"La Terre et l'humanité." *Annales des voyages* . . ., vol. 199 (Année 1868, vol. 3, July 1868), 5-44. [Extract from *La Terre*, vol. 2, which was then in preparation]

"L'Election présidentielle de la Plata et la guerre du Paraguay." *Revue des Deux Mondes*, vol. 76 (15 August 1868), 891-910.

1868-   *La Terre. Description des phénomènes de la vie du globe.* Paris:
1869   Hachette.
         Vol.   I   *Les Continents.* 1868.
         Vol.   II   *L'Océan-l'atmosphère-la vie.* 1869.

1869   With Elie Reclus. "Introduction." Pp. v-clxxxviii of Adolphe
         Joanne, *Dictionnaire géographique, administratif, postal, statistique, ar-
         chéologique, etc. de la France, de l'Algérie et des Colonies,* vol. 1.
         2nd ed. Paris: Hachette.

         "Les Voies de communication." Pp. 159-168 of *Almanach de la
         coopération* for 1869 (reprint in Bibliothèque Nationale.)

         Histoire d'un ruisseau. Paris: Bibliothèque d'éducation et de récréa-
         tion, J. Hetzel et Cie, n.d.

         "La Géographie." Pp. 109-112 of *Almanach de l'encyclopédie
         générale, Première année, 1869.* Paris: Librairie du Passage Euro-
         péen, Weil et Bloch.

1870   *Nice, Cannes, Antibes, Monaco, Menton, San-Rémo.* Paris:
         Hachette.

1870-   *Les Phénomènes terrestres.* Paris: Hachette.
1872   Vol.   I   *Les continents.* 1870.
         Vol.   II   *Les mers et les météores.* 1872.

1872-   Series of 25 unsigned articles, all but one titled "Géographie
1875   générale," in *La République Française* (Paris), 15 February 1872 to 8
         January 1875.

1873   "Les Pluies de la Suisse." *Bulletin de la Société de géographie*, 6th
         series, vol. 5 (January 1873), 88-91.

         "Notice sur les lacs des Alpes." *Bulletin de la Société de géographie*,
         6th series, vol. 5 (February 1873), 185-187.

         "Note relative à l'histoire de la mer d'Aral." *Bulletin de la Société de
         géographie*, 6th series, vol. 6 (August 1873), 113-118.

"Réponses aux observations précédentes" (following "A propos de la mer d'Aral" by Ali Suavi). *Bulletin de la Société de géographie,* 6th series, vol. 6 (November 1873), 533-536.

1874　"Extrait d'une lettre de M. Elisée Reclus au président de la Société." *Bulletin de la Société de géographie,* 6th series, vol. 7 (April 1874), 421-425.

"Voyage aux régions minières de la Transylvanie occidentale." *Le Tour du Monde,* vol. 28 (1874), 1-48.

1875　"Le Bosphore et la Mer Noire." *Le Globe* (Geneva),vol. 14 (1875), Mémoires, 19-35.

1876-　*Nouvelle géographie universelle. La Terre et les hommes.* Paris:
1894　Hachette.

Vol.	I	*L'Europe méridionale.* 1876.
	II	*La France.* 1877.
	III	*L'Europe centrale.* 1878.
	IV	*L'Europe du Nord-Ouest.* 1879.
	V	*L'Europe scandinave et russe.* 1880.
	VI	*L'Asie russe.* 1881.
	VII	*L'Asie orientale.* 1882.
	VIII	*L'Inde et l'Indo-Chine.* 1883.
	IX	*L'Asie antérieure.* 1884.
	X	*L'Afrique septentrionale,* First part, *Bassin du Nil.* 1885.
	XI	*L'Afrique septentrionale,* Second part, *Tripolitaine, Tunisie, Algérie, Maroc, Sahara.* 1886.
	XII	*L'Afrique occidentale.* 1887.
	XIII	*L'Afrique méridionale.* 1888.
	XIV	*Océan et terres océaniques.* 1889.
	XV	*Amérique boréale.* 1890.
	XVI	*Les Etats-Unis.* 1892.
	XVII	*Indes occidentales.* 1891.

(N.B. Reclus delayed the publication of Vol. XVI because he wanted to get more information about the United States. This accounts for the numbering and dating of Vols. XVI and XVII.)

	XVIII	*Amérique du sud. Les régions andines.* 1893.
	XIX	*Amérique du sud. L'Amazonie et La Plata.* 1894.

1877 "La Grève d'Amérique." *Le Travailleur*, vol. 1, no. 5 (September 1877), 6-16.

1878 "L'Evolution légale et l'anarchie." *Le Travailleur*, vol. 2 (January 1878), 7-14.

"La Passe du Sud et le port Eads dans le delta mississippien." *Revue lyonnaise de géographie*, vol. 1, no. 3 (12 January 1878), 145-149.

"L'Internationale et les chinois." *Le Travailleur*, vol. 2, no. 3 (March-April 1878), 22-31.

1880 "Ouvrier, prends la machine! Prends la terre, paysan!" *Le Révolté*, 1st year, no. 25 (24 January 1880), 1.

"Evolution et révolution." *Le Révolté*, 1st year, no. 27 (21 February 1880), 1-2.

Histoire d'une montagne. Paris: Bibliothèque d'éducation et de récréation, J. Hetzel et Cie, n. d.

"Rivers." Pp. 1651-1655 of *Johnson's New Universal Cyclopaedia*, ed. by Frederick A. P. Barnard and Arnold Guyot, vol. 3. New York: Alvin J. Johnson & Son.

1882 "L'Anarchie et le suffrage universel." *Le Révolté*, 3rd year, no. 24 (21 January 1882), 1-2.

1883 "Le Gouvernement et la morale." *Le Révolté*, 4th year, no. 23 (6 January 1883), 1.

1884 "Anarchy: By an Anarchist." *The Contemporary Review*, vol. 45, no. 5 (May 1884), 627-641.

1884- "Les Produits de la terre." 5-part article published in *Le Révolté*,
1885 6th year, no. 20 (23 November-6 December 1884) through no. 26 (15-28 February 1885). Sometimes attributed to Elisée Reclus, but he was apparently not the author, according to Jean Maitron, *Le Mouvement anarchiste en France* (Paris: Maspero, 1975), II, 407n.

1885   "Notes sur les Tuileries" (Letter from Reclus to Charles Normand, 24 September 1879). *Bulletin de la Société des amis des monuments parisiens,* vol. 1, no. 1 (1885), 15-17.

"Une lettre d'Elisée Reclus." *Le Révolté,* 2nd series, 1st year, no. 13 (11-24 October 1885), 1.

1887   "Scandinaves (Etats).—Suède, Norvège, Danemark." Pp. 1991-1995 of *Dictionnaire de pédagogie et d'instruction primaire,* ed. by Ferdinand Buisson, part 2, vol. 2. Paris: Hachette.

Review of Ludovic de Campou, *Un empire qui croule, le Maroc contemporaine. Bulletin de la Société neuchâteloise de géographie,* vol. 3 (1887), 138-140.

"Les Produits de l'industrie." *Le Révolté,* 8th year, no. 45 (26 February-4 March 1887), 1; no. 47 (12-18 March 1887), 1; no. 49 (26 March-1 April 1887), 1.

"La Richesse et la misère." 6-part article published in *Le Révolté/ La Révolte,* from *Le Révolté,* 9th year, no. 12 (25 June-1 July 1887) to *La Révolte,* 1st year, no. 8 (5-11 November 1887).

1889   "L'Evolution de la morale. Le Vol et les voleurs." *La Révolte,* 2nd year, no. 22 (10-16 February 1889), 1-2.

"Quelques notes sur la propriété." *La Société nouvelle,* 5th year, tome 1, no. 51 (31 March 1889), 322-329.

"Préface." Pp. v-xxviii of *La Civilisation et les grands fleuves historiques* by Léon Metchnikoff. Paris: Hachette.

1889-   "A propos d'une carte statistique" (Commentary on map of Paris
1890   and its suburbs by Charles Perron). *Bulletin de la Société neuchâteloise de géographie,* vol. 5 (1889-1890), 122-124.

1892   "Préface." Pp. v-xv of Peter Kropotkin, *La Conquête du pain.* (Bibliothèque sociologique) Paris: P.–V. Stock.

1894   "Hégémonie de l'Europe." *La Société nouvelle,* 10th year, tome 1, no. 112 (April 1894), 433-443.

"L'Idéal et la jeunesse." *La Société nouvelle*, 10th year, tome 1, no. 114 (June 1894), 721-731.

Nouvelle géographie universelle. La Terre et les hommes. Tableaux statistiques de tous les états comparés. Années 1890 à 1893. Paris: Hachette.

"East and West." *The Contemporary Review*, vol. 66, no. 346 (October 1894), 475-487.

"Quelques mots d'histoire." *La Société nouvelle*, 10th year, tome 2, no. 119 (November 1894), 489-494.

1895   "The Evolution of Cities." *The Contemporary Review*, vol. 67, no. 2 (February 1895), 246-264.

"Russia, Mongolia, and China." *The Contemporary Review*, vol. 67, no. 5 (May 1895), 617-624.

"L'opinione de E. Reclus sull'eventuale emigrazione cinese in Europa." *Bollettino della Società geografica italiana*, 3rd series, vol. 8, no. 6 (June 1895), 174-175.

"La Cité du bon accord." Pp. 103-106 in *The Evergreen, A Northern Seasonal*, Part 2, *The Book of Autumn*. Edinburgh: Patrick Geddes and Colleagues.

"Discours de M. Elisée Reclus." Pp. 3-13 of *Séance solennelle de rentrée du 22 octobre 1895* (Université Nouvelle de Bruxelles). Brussels: Imprimerie Veuve Ferdinand Larcier.

Projet de construction d'un globe terrestre à l'échelle du cent-millième. Paris: Edition de la Société nouvelle.

"Recent Books on the United States." *Geographical Journal*, vol. 6, no. 5 (November 1895), 448-453. [Review]

1896   "Projet de construction d'un globe terrestre à l'échelle du 100,000^e." Pp. 625-636 of *Report of the Sixth International Geographical Congress, Held in London, 1895*. London: John Murray.

With Elie Reclus. "Renouveau d'une cité." *La Société nouvelle*, 12th year, tome 1, no. 138 (June 1896), 752-758.

"The Progress of Mankind." *The Contemporary Review*, vol. 70 (December 1896), 761-783.

1896-
1897
With Georges Guyou (Paul Reclus). "D'un Atlas à échelle uniforme." *Bulletin de la Société neuchâteloise de géographie*, vol. 9 (1896-1897), 159-164.

1897
"Quelques mots sur la révolution bouddhique." *L'Humanité nouvelle*, vol. 1 (1897), 139-145.

"Préface." Pp. v-viii of *Bibliographie de l'anarchie* by Max Nettlau. (Bibliothèque des "Temps Nouveaux," Année 1897, no. 8) Brussels: Bibliothèque des "Temps Nouveaux"; Paris: P.–V. Stock.

"Préface." Pp. v-xi of *Le Socialisme en danger* by F. Domela Nieuwenhuis. (Bibliothèque sociologique, 15) Paris: P.–V. Stock.

1898
"Attila de Gerando." *Revue de géographie*, vol. 42, no. 1 (January 1898), 1-4.

"Pages de sociologie préhistorique." *L'Humanité nouvelle*, vol. 2, no. 8 (February 1898), 129-143.

"The Vivisection of China." *The Atlantic Monthly*, vol. 82, no. 491 (September 1898), 329-338.

"A Great Globe." *Geographical Journal*, vol. 12, no. 4 (October 1898), 401-406.

Review of Edmond Demolins, *Les Français d'aujourd'hui* (Paris: Firmin Didot, 1898). *L'Humanité nouvelle*, vol. 3, no. 17 (November 1898), 628-632.

"L'Extrême-Orient." *Bulletin de la Société royale de géographie d'Anvers*, vol. 22 (1898), 143-145.

L'Evolution, la révolution et l'idéal anarchique. (Bibliothèque sociologique, 19). Paris: P.–V. Stock.

BIBLIOGRAPHY

1899   Review of Anonymous, *Les Communes mixtes et le gouvernement des indigènes en Algérie* (Paris: Challamel, 1897). *L'Humanité nouvelle*, vol. 4, no. 19 (January 1899), 118-119.

Review of Arsène Dumont, *Natalité et démocratie* (Paris: Schleicher frères, 1898). *L'Humanité nouvelle*, vol. 4, no. 20 (February 1899), 260-261.

Review of René Gonnard, *La Dépopulation en France* (Lyon: A.–H. Storck, 1898). *L'Humanité nouvelle*, vol. 4, no. 21 (10 March 1899), 377.

Review of Pierre Foncin, *Les Pays de France. Projet de fédéralisme administratif* (Paris: A. Colin, 1898). *L'Humanité nouvelle*, vol. 4, no. 21 (10 March 1899), 381-382.

Review of Sven Hedin, *Trois ans de lutte aux déserts d'Asie* (trans. by Charles Rabot) (Paris: Hachette, 1899). *L'Humanité nouvelle*, vol. 4, no. 21 (10 March 1899), 394-395.

Review of Edouard Foa, *Le Dahomey* (Paris: A. Hennuyer, 1895). *L'Humanité nouvelle*, vol. 4, no. 22 (10 April 1899), 505-506.

"The International Routes of Asia." *The Independent*, vol. 51 (4 May 1899), 1210-1215.

Review of Pierre d'Espagnat, *Jours de Guinée* (Paris: Perrin, 1899). *L'Humanité nouvelle*, vol. 4, no. 23 (10 May 1899), 626.

Review of Martin A. S. Hume, *Spain, Its Greatness and Decay (1479-1788)* (Cambridge: Cambridge University Press, 1898). *L'Humanité nouvelle*, vol. 4, no. 23 (10 May 1899), 640-641.

Review of A. J. Wauters, *L'Etat indépendant du Congo* (Brussels: Falk fils, 1899). *L'Humanité nouvelle*, vol. 4, no. 24 (10 June 1899), 750-751.

Review of E. W. Sikes, *The Transition of North Carolina from Colony to Commonwealth* (Baltimore: The Johns Hopkins Press, 1898). *L'Humanité nouvelle*, vol. 5, no. 25 (10 July 1899), 126-127.

[177]

Review of Ramon Reyes Lala, *The Philippine Islands* (New York: Continental Publishing Company, 1899); Frederic Noa, *The Pearl of the Antilles* (New York: The Knickerbocker Press, 1898); and Thomas Campbell-Copeland, *American Colonial Handbook* (New York: Funk and Wagnalls, 1899). *L'Humanité nouvelle*, vol. 5, no. 26 (10 August 1899), 245.

Review of Léopold de Saussure, *Psychologie de la colonisation française dans ses rapports avec les sociétés indigènes* (Paris: Félix Alcan, 1899). *L'Humanité nouvelle*, vol. 5, no. 26 (10 August 1899), 246-248.

Review of Roberto J. Payró, *La Australia Argentina* (Buenos Aires: La Nación, 1898). *L'Humanité nouvelle*, vol. 5, no. 26 (10 August 1899), 248-249.

Review of Eugène Aubin, *Les Anglais en Inde et en Egypte* (Paris: A. Colin, 1899). *L'Humanité nouvelle*, vol. 5, no. 27 (10 September 1899), 379-380.

Review of J. de Saint-Maurice Joleaud-Barral, *La Colonisation française en Annam et au Tonkin* (Paris: Plon, 1899). *L'Humanité nouvelle*, vol. 5, no. 28 (10 October 1899), 511-512.

Review of Lucien Jottrand, *Croquis du Nord, Nordland, Finmark, Spitzberg* (Paris: Lemoigne, 1898). *L'Humanité nouvelle*, vol. 5, no. 29 (10 November 1899), 638.

"On a Proposed Great Globe" (Abstract). *Report of the Sixty-Eighth Meeting of the British Association for the Advancement of Science, Held at Bristol in September 1898.* London: John Murray. Page 945.

"La Perse." *Bulletin of the Société neuchâteloise de géographie*, vol. 11 (1899), 27-62.

1900   Review of Willy Bambus, *Palästina, land und leute. Reiseschilderungen* (Berlin: S. Cronbach, 1898). *L'Humanité nouvelle*, vol. 6, no. 31 (January 1900), 119-120.

Review of Pierre de Barthélemy, *En l'Indo-Chine* . . . (Paris: Plon, 1899). *L'Humanité nouvelle*, vol. 6, no. 31 (January 1900), 120.

Review of Alphonse de Haulleville, *Les Aptitudes colonisatrices des Belges et la question coloniale en Belgique* (Brussels: J. Lebègue et Cie., 1898). *L'Humanité nouvelle*, vol. 6, no. 33 (March 1900), 369-370.

Review of Th. Bentzon (pseudonym of Thérèse Blanc), *Notes en· voyage. Nouvelle-France et Nouvelle-Angleterre* (Paris: Calmann Lévy, 1899). *L'Humanité nouvelle*, vol. 6, no. 35 (May 1900), 617.

Review of R. Meldi, *La Colonia Eritrea* (Parma: Battei, 1899). *L'Humanité nouvelle*, vol. 6, no. 35 (May 1900), 620.

"Les Colonies anarchistes." *Les Temps Nouveaux*, 6th year, no. 11 (7-13 July 1900), 1-2.

"La Chine et la diplomatie européenne." *L'Humanité nouvelle*, vol. 7, no. 39 (September 1900), 257-270.

With George Guyou (Paul Reclus). "L'Anarchie et l'église." *Les Temps Nouveaux*, no. 20 (1900).

"La Phénicie et les phéniciens." *Bulletin de la Société neuchâteloise de géographie*, vol. 12 (1900), 261-274.

1901   "On Vegetarianism." *The Humane Review*, vol. 1, no. 4 (January 1901), 316-324. Published in French as "A propos du végétarisme," *La Réforme alimentaire*, March 1901, 37-45, and reprinted as a 15-page· pamphlet, *Le Végétarisme* (Brussels: Le Naturiste, 1906).

"The Teaching of Geography. Globes, Discs, and Reliefs." *Scottish Geographical Magazine*, vol. 17, no. 8 (August 1901), 393-399.

With George P. Reclus-Guyou (Paul Reclus). "On a One-Scaled Atlas." *Bulletin of the American Bureau of Geography*, vol. 2, no. 3 (September 1901), 199-204.

L'Enseignement de la géographie. Globes, disques globulaires et reliefs. Université Nouvelle, Institut géographique de Bruxelles, Publication No. 5, 1901.

L'Afrique australe. Edited by Onésime Reclus. Paris: Hachette.

1902 "Annexo. Carta de M. Elisée Reclus" (Letter following article by Paul Choffat, J. Barbosa Bettencourt, and E. de Vasconcellos, "Açores. A que parte do mundo devem pertencer?") *Boletim da Sociedade de Geographia de Lisboa,* vol. 20, nos. 1-6 (January-June 1902), 365-366.

Review of W. P. Livingstone, *Black Jamaica. A Study in Evolution* (London: Sampson Low, 1899). *L'Humanité nouvelle,* vol. 8, no. 45 (October 1902), 112-114.

Review of Arthur Buies, *La Province de Québec* (Québec: Département de l'agriculture de la province de Québec, 1900). *L'Humanité nouvelle,* vol. 8, no. 46 (November 1902), 233-234.

"Préface." Pp. 5-10 of *Fin de vie (notes et souvenirs)* by Eugène Noel. Rouen: Imprimerie Julien Lecerf.

With Onésime Reclus. *L'Empire du milieu. Le Climat, le sol, les races, la richesse de la Chine.* Paris: Hachette.

"Préface." Pp. i-iv of *Voyage de la "Belgica," Quinze mois dans l'Antarctique* by Adrien de Gerlache. Paris: Hachette; Brussels: G. Lebèque et Cie.

"Der 'Disque Globulaire'." *Zeitschrift der Gesellschaft für Erdkunde zu Berlin,* 1902, 57-59.

1903 Review of *Atlas de Finlande* (Helsingfors: Société de géographie de Finlande, 1899). *L'Humanité nouvelle,* vol. 9, no. 47 (May 1903), 107-110.

Review of Camille Dreyfus, *A la Côte d'Ivoire, six mois dans l'Attié (un Transvaal français)* (Paris: L.–H. May, 1900). *L'Humanité nouvelle,* vol. 9, no. 49 (July 1903), 323-324.

Review of Anonymous, *Les Chemins de fer du grand-duché de Finlande* (Helsingfors: Imprimerie du gouvernement, n.d.). *L'Humanité nouvelle*, vol. 9, no. 49 (July 1903), 324.

Review of J. W. Powell, *Annual Report of the Bureau of American Ethnology (1897-1898)*, part 1 (Washington, 1900). *L'Humanité nouvelle*, vol. 9, no. 50 (August 1903), 442-445.

"On Spherical Maps and Reliefs." *Geographical Journal*, vol. 22, no. 3 (September 1903), 290-293.

"Le Panslavisme et l'unité russe." *La Revue*, vol. 47 (1 November 1903), 273-284.

"Chronique géographique." *La Revue*, vol. 47 (15 November 1903), 520-521.

Review of Paul Pelet, *Atlas des colonies françaises* (Paris: Armand Colin, 1902). *La Revue*, vol. 47 (1 December 1903), 648.

"Mouvement géographique." *La Revue*, vol. 47 (15 December 1903), 779-781.

"L'Enseignement de la géographie." Reprinted from *Bulletin de la Société belge d'astronomie*, no. 1 (1903), 3-9.

"Proposition de dresser une carte authentique des volcans." Reprint from *Bulletin de la Société belge d'astronomie*, no. 11 (1903).

1904   "Mouvement géographique." *La Revue*, vol. 48 (15 January 1904), 253-255.

"Le Patriotisme est-il incompatible avec l'amour de l'humanité? Enquête." *La Revue*, vol. 48 (15 January 1904), 169-170.

"Mouvement géographique." *La Revue*, vol. 48 (15 February 1904), 517-518.

"A propos de la guerre d'Extrême-Orient." *La Revue*, vol. 48 (1 April 1904), 304-308.

"Mouvement géographique." *La Revue*, vol. 49 (1 May 1904), 104-105.

Review of Napoleone Colajanni, *Razze inferiori e razze superiori; o, Latini e Anglo-Sassoni* (Rome: Presso la Rivista popolare illustrata, 1903). *La Revue*, vol. 49 (15 May 1904), 225-226.

Review of Gabriel Giroud, *Population et subsistances, essai d'arithmétique éconömique* (Paris: C. Reinwald, 1904). *La Revue*, vol. 51 (1 September 1904), 100-101.

"Les grandes voies historiques" (Résumé of Reclus's lectures by E. Cammaerts). *Bulletin de la Société royale belge de géographie*, vol. 28 (1904), 5-15.

"Aperçu géographique." Pp. 35-80 of *Le Mexique au debut du XXe siècle*, vol. 1. Paris: Librairie Ch. Delagrave, n.d.

1905   "Une voix d'Häiti." *La Revue*, vol. 55 (1 June 1905), 393-395.

"Nouvelle proposition pour la suppression de l'Ere chrétienne." *Les Temps Nouveaux*, vol. 11, no. 1 (6 May 1905), 1-2.

Elie Reclus, 1827-1904. Paris: L'Emancipatrice (imprimerie communiste), n.d.

Introduction. Introductory volume to *Dictionnaire géographique et administratif de la France*, edited by Paul Joanne (7 vols., A-Z, published 1890-1905). Paris: Hachette.

1905-   *L'Homme et la terre.* 6 vols. Paris: Librairie Universelle.
1908 Vol. I   *Les Ancêtres. Histoire ancienne.*   1905.
       II   *Histoire ancienne.*   1905.
       III   *Histoire ancienne. Histoire moderne.*   1905.
       IV   *Histoire moderne (cont.)*   1905.
       V   *Histoire moderne (cont.)—Histoire contemporaine.*   1905.
       VI   *Histoire contemporaine.*   1908.

Second edition, abridged to three volumes and edited by G. Goujon, A. Perpillou, and Paul Reclus. Paris: Albin Michel, Editeur, [1930]-1931.

1906-   *Les Volcans de la terre.* 3 parts. Brussels: Société belge d'as-
1910   tronomie, de météorologie et du physique du globe.

         Part 1   Asie antérieure.   1906.

             2   Mediterranée et Europe centrale.   1908.

             3   Italie et Sicile (cont.).   1910.

1911-   Correspondance. 3 vols.
1925   Vol. 1   Paris: Librairie Schleicher Frères, 1911.

                     Edited by Louise Dumesnil.

             2   Paris: Librairie Schleicher Frères, 1911.

                     Edited by Louise Dumesnil.

             3   Paris: Alfred Costes, Editeur, 1925.

                     Edited by Paul Reclus.

Index

The relationship of members of his family to Elisée Reclus is indicated within parentheses.

INDEX

INDEX